Doing Honest Work in College
How to Prepare Citations, Avoid Plagiarism,
and Achieve Real Academic Success

Third Edition

（美）查尔斯·利普森（Charles Lipson） 著

龚伟峰 译

学术诚信与引用规范
如何避免剽窃，获得学术成功

（原著第三版）

  化学工业出版社

·北京·

内 容 简 介

本书是一本学术规范工具书,分学术诚信和引用格式规范两部分。学术诚信部分主要介绍了学术诚信的基本原则,日常学习和科研中如何培养学术规范的意识,如何养成优秀的笔记习惯,以及怎样合理引用和转述。引用格式规范部分主要介绍了不同学科领域的文献引用规范,通过表格的方式列举了各种格式引用要点及文献引用举例,包括芝加哥(或杜拉宾式)引用格式,MLA格式,APA格式,CSE格式,AMA格式,ACS格式,以及物理学、天体物理学和天文学,数学、计算机科学和工程学的引用格式。另外本书还对引用格式的常见问题进行了解答。

不同于官方手册,本书没有烦琐的规则和条目,而是采用举例的方式,列举了每种文献引用要点,读者只需根据文献类型检索即可了解到相应的引用方法。

本书可供高等院校本科生、研究生参考使用。

Doing Honest Work in College: How to Prepare Citations, Avoid Plagiarism, and Achieve Real Academic Success, third edition/by Charles Lipson
ISBN 9780226430744
© 2004, 2008, 2018 by Charles Lipson. All rights reserved.
Authorized translation by The University of Chicago Press, Chicago, Illinois, U.S.A
本书中文简体字版由 The University of Chicago Press 授权化学工业出版社有限公司独家出版发行。
本书仅限在中国内地(大陆)销售,不得销往中国香港、澳门和台湾地区。未经许可,不得以任何方式复制或抄袭本书的任何部分,违者必究。

北京市版权局著作权合同登记号:01-2022-5257

图书在版编目(CIP)数据

学术诚信与引用规范:如何避免剽窃,获得学术成功/(美)查尔斯·利普森(Charles Lipson)著;龚伟峰译.—北京:化学工业出版社,2022.9
书名原文:Doing Honest Work in College: How to Prepare Citations, Avoid Plagiarism, and Achieve Real Academic Success
ISBN 978-7-122-41858-6

Ⅰ.①学… Ⅱ.①查… ②龚… Ⅲ.①学术研究-道德规范-研究 Ⅳ.①G30

中国版本图书馆CIP数据核字(2022)第146621号

责任编辑:韩霄翠
文字编辑:刘 璐
责任校对:宋 玮
装帧设计:史利平

出版发行:化学工业出版社
　　　　(北京市东城区青年湖南街13号 邮政编码100011)
印　装:大厂聚鑫印刷有限责任公司
787mm×1092mm 1/16 印张10¼ 字数216千字
2022年11月北京第1版第1次印刷

购书咨询:010-64518888
售后服务:010-64518899
网　址:http://www.cip.com.cn
凡购买本书,如有缺损质量问题,本社销售中心负责调换。

定　价:58.00元　　　　　　　　　　版权所有　违者必究

译者前言

如果你走进一间教室，请教室里的学生描述何为"剽窃"，你将收获许多条定义，这些定义风格迥异，其中一些甚至相互矛盾。国内外的许多研究者都对此做过调研，我在自己的课堂上也做过类似的互动，那条几乎所有人都倡导的金科玉律——不许剽窃——在不同的人心中，却代表着极为不同的行事方式。而如果"剽窃"的概念不能明确，"学术诚信"也就难以真正实现。

从进入学校开始，老师们就反复告诫我们：不许抄作业。这条指令是清晰的吗？如果我们使用了参考书，借鉴了其中的想法，算不算抄作业？如果我们需要完成的是一项小组任务，有人尽力、有人躺平，而大家交流探讨的过程又算不算抄作业？同样的情况放到实验室的科研工作中呢，为了让合作切实得以开展，我们又该遵循哪些行为准则？

一篇论文涉嫌"剽窃"，这是我们判定的"果"，而"剽窃"行为或许早在阅读文献之时就悄然发生了，有些甚至是无意之举——如果此刻能有一位经验丰富的作者，用简单、风趣的语言把这些问题都讲清楚，帮助你少走弯路，你会愿意听听他的建议吗？

Lipson 教授自 2004 年就开始围绕上述话题进行写作，你现在翻开的这本书已经是 *Doing Honest Work in College：How to Prepare Citations, Avoid Plagiarism, and Achieve Real Academic Success* 的第 3 版。如果你想找到上述问题的答案，并且希望了解不同领域的专家、导师对这些问题的看法，请通读本书的第一部分（第 1 章至第 4 章）。针对学术诚信、小组作业、合理引用等耳熟能详的话题，有许多新鲜的视角等待你发掘。

本书的第二部分"引用格式快速入门"，则服务于不同的研究领域，阅读时，可先完成第 5 章"引用格式概述"，再根据自身需求，跳跃到相关章节深入学习。我的个人感受是：Lipson 教授在解读各种引用格式的过程中，形成了一套自己的独特思路，这比直接阅读官方手册要高效许多。

Lipson 教授希望他在书中介绍的各项原则能够维护学术公平，促进真才实学，而这也是我翻译这本书的最大动力所在。感谢化学工业出版社编辑的细心编校。由于译者水平有限，不足之处在所难免，敬请读者批评指正。

龚伟峰

2022 年 6 月

本书常用缩写

ACS   American Chemical Society 美国化学学会
AIP   American Institute of Physics 美国物理联合会
AMA   American Medical Association 美国医学会
AMS   American Mathematical Society 美国数学学会
APA   American Psychological Association 美国心理学会
ASCE   American Society of Civil Engineers 美国土木工程师协会
Chicago   *The Chicago Manual of Style* 《芝加哥手册》
CSE   Council of Science Editors 国际科学编辑委员会
IEEE   Institute of Electrical and Electronics Engineers 电气电子工程师学会
MLA   Modern Language Association 美国现代语言学会

目 录

第一部分 ▶ 学术诚信   1

1 学术诚信三原则   2

2 学术诚信：从课堂到考试   4
    2.1 阅读任务   5
    2.2 考试   5
    2.3 论文   7
    2.4 使用网络资源   9
    2.5 语言课   11
    2.6 小组作业   12
    2.7 加入实验室   17
    2.8 课堂参与   22
    2.9 低分申诉   23
    2.10 荣誉行为守则   25

3 优秀笔记养成秘籍   26
    3.1 何为优秀的笔记   26
    3.2 听课笔记和读书笔记   28
    3.3 复习和改进你的笔记   30
    3.4 用字母"Q"标记法做笔记   30
    3.5 使用英文省略号缩短引文   32
    3.6 使用方括号为引文增加可读性   33
    3.7 引文内部含有另一层引文   33
    3.8 字母"Q"标记法的综合应用   33

4 剽窃与学术诚信   35
    4.1 合理引用他人成果   36
    4.2 合理引用网络资源   36
    4.3 合理地引用和转述   37
    4.4 如何转述   38
    4.5 剽窃他人的观点   41
    4.6 歪曲他人的观点   41
    4.7 结论：转述和引用的正确方式   41

第二部分
引用格式快速入门   43

5 引用格式概述   44
    5.1 悬挂缩进   47
    5.2 到哪里获取更多信息   48
    5.3 开始具体操作   48

6 芝加哥（或杜拉宾式）引用格式   50
    6.1 完整注释、简略注释和参考文献   50
    6.2 芝加哥引用格式的常见问题   74

7 MLA 格式：人文学科   76
    7.1 MLA 格式的用法   76
    7.2 MLA 格式的常见问题   92

8 APA 格式：社会科学、教育学和商学   94

9 CSE 格式：生物科学   112
    9.1 CSE 格式的三种引注方式   112
    9.2 参考文献列表的格式   114
    9.3 文内引注和参考文献列表的引用举例   115

10 AMA 格式：生物医学、医学、护理学和牙医学   121

11 ACS 格式：化学   126

12 物理学、天体物理学和天文学的引用格式   131
    12.1 物理学常用的 AIP 格式   131
    12.2 天体物理学和天文学的引用格式   132

13 数学、计算机科学和工程学的引用格式   135
    13.1 数学和计算机科学   135
    13.2 数学论文的文本样式   138
    13.3 计算机科学：在编程中引用源代码   139
    13.4 工程学的 IEEE 引用格式   139
    13.5 工程学的 ASCE 引用格式   141

14 各类引用格式常见问题解答   144
    14.1 哪些内容需要引用？   144
    14.2 如何应对引用中的难题？   145
    14.3 参考文献   147
    14.4 直接引用   149
    14.5 网络资源   150

致   谢   153

索   引   154

第一部分
学术诚信

1

学术诚信三原则

学术诚信可以归结为三项简单而有力的原则：

① 当你宣称自己完成了某项工作时，你实际上确实做了。

② 如果参考了他人成果，需要如实引注。直接引用时，要明确标注、写清出处、勿做修改。

③ 呈现研究资料时，要做到公正合理、真实客观。这里所说的研究资料包括但不限于各类数据、文件以及其他研究者的著作。

记住并遵守上述原则并不困难，它们是实现学术诚信的基石，适用于你参与的所有课程、实验、考试和论文。无论是英文论文、化学实验，还是程序设计、建筑图纸，都受到三项原则的约束。大学中的全体成员也都要遵照执行，无论是大一新生，还是学校教师。

当然，每所大学有其相应的行为准则，每个课堂也会针对不同任务提出具体要求。我将在第 2 章详细讨论这些内容，并重点说明如何在论文、实验、小组作业和考试中遵守相关规定。

此外，我会探讨在作业、测试中合理使用网络资源的方式：到底能否借助电脑或手机；如何引用网站、社交媒体、播客、著作、文章、诗歌、电影等来源的信息。可以说，你手中的这本小册子横跨计算机科学、视觉艺术等多个领域，既能帮助你避免剽窃，还能应对绝大多数有关引用规范的问题。

再说到科技领域，尤其是研究组和实验室中的诚信原则。一些人对此感到费解——因为有时候需要同伴参与，有时候需要独立工作。其中又有哪些事必须自己做，他人无法代劳呢？我会为你转述实验室负责人的建议，他们阐述了应该如何充分发挥团队的力量，并且避开潜在的问题。

本书为所有问题提供的回答，都基于与大学教务长❶的交流。他们每天处理学术诚信

---

❶ 国外高校设置的高级管理职务，主管全校教学工作。——译者注

问题，对此相当了解。相信我，在这本书里阅读教务长的建议，远比犯了错听他们当面教导要好得多——你一定不想走到那一步。

最重要的是听从教师对不同任务的要求，一旦有不清楚的地方，记得及时询问，然后确保遵守学术诚信三项基本原则：如果你说你做了某项工作，那你必须确实做了；如果引用了他人文字或成果，你必须公开承认；呈现研究资料时，公正合理、真实客观。千万不要伪造数据、隐藏不好的结果或偷窃他人成果。

最基本的原则是：诚实公正地呈现你所做的工作。不得歪曲自己或他人的研究发现。不得错误表述他人观点。参考他人成果完成论文或答卷时，不得谎称个人原创。对于他人的贡献和付出，要给予充分认可。写论文时，引用文献、注明出处。参加考试时，独立完成，绝不作弊。

只要你照此行事，就不会违反学校的规则，也就能做到真正的学术诚信，从而扎实地学习每一份资料，诚实地获取每一点成绩。

下面，我们来聊聊具体如何操作。

2

学术诚信：从课堂到考试

在大学里坚守学术诚信，意味着要做哪些事呢？本章就一些日常事务展开探讨，并提供实用的解决方案。首先是阅读任务和测试。从开课第一天起，老师就会要求你读书、读文章，或许还有看视频、做实验以及完成小组任务。你要搞清楚老师期望你达到的目标是什么。搞清楚目标对于研讨课、各种讨论环节，甚至外语等专项课程都至关重要。我会为这些活动提供具体建议，帮你找准方向。

开课一段时间之后你将迎来期中考试，无论地点在家还是在教室，所有考试都有一个基本原则：独立完成，严禁抄袭。在此基础上，不同形式的测验还有各自的规则。我会针对不同的规则做出说明，并提醒你在不明白某些规则时该怎么办。

然后是大学教育中非常重要的一项任务：写论文。论文写作需要参考他人成果，并将它们与你的原创观点有机结合。注意事项则是要有清晰的笔记和合理的引用。这也是本章和下一章的重点，我将介绍做笔记的方法，帮助你在笔记环节将自己和他人的观点明确区分开来；还将讲解参考他人著作时，如何引用和转述各类印刷出版或在线出版的文献。

阅读、考试、论文等都属于个人项目，还有一些则是团队项目，例如在化学、生物学领域，你需要跟实验室的同事紧密合作；在统计学、计算机领域，你要加入某个课题组。在这种情况下，哪些工作由团队完成，哪些又必须各自完成呢？我会在后文予以解答。

另外要向你解释的是：如何面对不够理想的成绩。只要你愿意，你永远可以寻求改进建议。老师们十分乐于帮助学生，尤其是当你希望改变又肯付诸行动的时候。如果你觉得自己的论文被评判得过于苛刻了，可以试着提出申诉，但是这里也有一定的基本规则——你不能在论文或试卷上做任何改动，且必须提供清晰、合理的申诉理由。详细信息在本章后面的部分呈现。

本章最后讨论大学的荣誉行为守则，其作用是鼓励研究者诚实做学问，并为个人行为承担责任。

第 2 章的内容旨在引导你从宏观上把握在大学里践行学术诚信的内涵，帮助你逐步应对可能遇到的问题，范围涵盖了阅读任务、考试、论文、小组作业、实验等，以及一些偶发问题的处理建议。我还会分享一些改善学习效果的窍门。当然，归根结底只为一件事：

诚实地学习。

2.1 阅读任务

老师们刚一开课就会布置阅读任务。偷懒不读是不对的,但不构成作弊。有的学生会到 SparkNotes 或 CliffsNotes 网站上找些概要来读,作为补充。或者干脆就只读概要,不读原著。如果你觉得这类摘要能帮到你,那么把它们作为补充资料是可以的,但因此而越过原著是偷工减料、自作聪明。你错过了原著本身的持久价值,错失了阅读、领悟它们的机会。至此,你尚未违反任何学术规范。

> **高效阅读的建议:**
>
> 拿到文章或非虚构类书籍,翻开第一页然后从头读到尾——这种做法并不可取。建议你先查看引言、结论和目录,获取阅读材料的全貌。对于一篇论文,从摘要和小标题开始浏览,然后读引言和结论。既然不是在读悬疑小说,就不用担心提前知道结论。做完这些,你会对阅读任务形成整体感知,接下来就可以更深入地阅读了。这也是老师们的阅读方式。

2.2 考试

常见的考试有两类:一类在考场,一类可以带回家。形式不同,规则也不同。我们依次来看。

(1) 考场考试

在教室中的测验,大部分都是闭卷考试。你不得查阅任何资料,包括手写笔记以及存储在电脑或移动设备上的数据。考试可能会需要你自备空白的蓝皮答题册并带入考场。以上是考场的默认规则,通常不会发生变化,请你严格遵守,除非监考人员明确告知新的规则。如果你对某场考试的规则存有疑问,及时向老师咨询。

同时,开考之前你不得接触试卷。或许你能猜到考试范围,也可能看过上一学期或学年的试题,但是你无权在当前试卷启用前查看其内容。将上午场次的考题泄露给下午场次的考生是作弊行为。

有时候老师会在课内进行开卷考试。学生可以翻书、查笔记,使用移动设备,甚至能临场用笔记本电脑录入答案。开卷考试的规则大致就是如此。允许查资料的好处是省去了

---

❶ 美国大学生考试用的答题本,即 blue books。——译者注

枯燥的复习和死记硬背，但这并不意味着开卷考试更加轻松。为了写出令人信服的答案，或在必要时迅速定位关键信息，学生仍要对相关的材料了如指掌。请注意：开卷不等于将网络上的内容直接搬运到你的试卷上。你可以浏览不同的资源，但答案要靠自己得出。原则是不变的，你不能把他人的作品当成自己的——对于考试、论文或实验，都是如此。

传闻某位老师曾宣布，在下一场开卷测试中，学生可以借助任何随身带入教室的东西完成考试。一个考生索性背起一名研究生，"随身携带"，入场答题。这位考生在钻空子方面确实足够努力，或许未来可以考虑律师行业。

实际上，你的考试得由你独立完成才行。不管是开卷还是闭卷，都不可以偷看他人的答卷，不得使用网络或电脑里预先下载好的答案，不许发信息求助，不能剽窃他人的语言或观点。最后，禁止把研究生扛进考场，体重再轻也不行。

考场考试的建议：

如无特殊说明，考场上的默认规则是闭卷，即在答题时不得使用书本、笔记、文章或线上资料。

考前跟同学结伴学习是可以的。大家一起切磋、琢磨答案对学习有利，前提是你积极参与其中。若只是坐在一旁当听众，效果会大打折扣，从这个角度讲，学习小组的规模小一些更合适。然而一旦进入考场，你就不能指望别人了，这是学术诚信的集中体现。

伍迪·艾伦就为此受到过教训，他说："大一期末考形而上学的时候，我偷窥了旁边男孩的灵魂……因为这次作弊纽约大学把我开除了。" ❶

（2）带回家考试

带回家考试一般是开卷，允许参考书本、文章、笔记和网络。为确保万无一失，每次考前不妨核实一下具体要求。

可以查阅现成的出版物和自己的笔记，但不能找其他人帮你，否则会构成作弊。随堂考试也是不允许向他人求助的。

同样，如果作答时吸纳了来自书本、论文或网络的观点，则要进行标注。尽管这是开卷考试，不是一篇科研论文，我依然建议你谨慎对待，合理引用（后面我会详细讲解该怎么做）。直接引用的部分要加引号，并写明出处。转写的部分，要用自己的语言表述，避免大量套用作者原话；同时，记得标注出处。

在符合引用规范的基础上，你的答卷必须是你的作品。你不能从网上整段地复制，把

---

❶ 引自影片《安妮·霍尔》，由伍迪·艾伦担任导演，伍迪·艾伦、马歇尔·布瑞克曼共同编剧。
The quote is from *Annie Hall*, directed by Woody Allen, written by Woody Allen and Marshall Brickman (MGM/UA Studios, 1977; DVD, Santa Monica: MGM Home Video, 2003).

它们粘贴到试卷或论文里,插入一两条参考文献,然后说这是你的作品。东拼西凑,复制粘贴,拿别人的果实应付差事,谈不上原创性。

事实上,考试和论文在这方面的规定是格外严格的。你不可以从网上、书上或其他任何地方照抄任何内容,哪怕是短短一句都不行——除非进行明确标注和合理引用。基本原则要记牢:只要参考了他人成果,就要引用;借用他人原话时,要公开、准确地引用它们。

> **带回家考试的建议:**
>
> 在家考试通常允许借助书本、文献、笔记和网络,但是建议在考前问清具体的规定。两件事坚决不能做:照抄答案和找人帮忙。考试须由你独立完成。间接引用的内容,要用自己的话转述,还要注明出处。如果是原封不动的引用,要把它们放在引号里,同时注明出处。

2.3 论文

论文基本上还是个人项目(我会在后文讨论小组作业,本节聚焦个体需要完成的工作)。你要独立完成阅读、研究和写作,并为结果负责。你可以跟其他人交流自己的课题,向老师、写作导师以及朋友咨询,甚至可以把草稿分享给他们,寻求反馈。但是没人可以代替你做研究、构建论文框架或完成写作。

你几乎总是在前人研究的基础上开展工作,人们也会据此来评估你的成果。所以,要挑最好的作品来读,还要带着批判的眼光去读,不能一味地赞同。高等教育的一个重要使命就在于此。练就思辨能力,你会变得见多识广、思虑周全。

这一节要讲的不是怎么把论文写完,而是更明确的一件事——你必须诚实地写作。原则很简单。第一,凡是参考他人著作的部分,你都要注明,也就是让读者知情。不管你是准备支持还是反驳所引用的文献,你都应该这么做。如果借用了他人观点,引用加标注。如果借用了他人原话,使用引号,同时注明来源。

第二,凡是直接引用都应该添加引号。但若是援引的部分过长,有许多句话,建议增加缩进,并取消引号。例如,读者看到下面这部分整块缩进的内容,就会知道是直接引用:

> Four score and seven years ago our fathers brought forth on this continent, a new nation, conceived in Liberty, and dedicated to the proposition that all men are created equal.
>
> Now we are engaged in a great civil war, testing whether that nation, or any nation so

conceived and so dedicated, can long endure.❶

不会有人觉得你打算剽窃《葛底斯堡演说》。不过，你依然需要注明出处。如果文献的出处本身就存有争议或是没什么名气，清晰的引注就格外重要了。无论你借用了文献中的数据、分析还是语句，读者都有权知道它们的确切来源。

换言之，你附上的参考文献（和参考文献的数量）不可以误导读者。要善于利用文献来加强论证，丰富观点，但是不能把文献当成自己原创性工作的替代品。

关于论文还有一条更重要的注意事项：不能在不同的课程提交相同的论文。其他类型的家庭作业也遵循这条原则。例如，你为经济学课写了一篇有关大萧条的论文，就不能把它再提交到美国史课程上。你得另写一篇，也就是说，不允许剽窃自己的作品。你可以继续以大萧条为主题，甚至选取一些相同的参考书目、文献，但成品必须跟上一篇有本质区别。（如果你确实希望在上一篇论文的基础上进行扩写，那么要告知当前课程的教师并取得许可。如果你希望把同一篇论文用于两门课，要向两位教师说明并取得许可。）

诚实写作的建议：

- 凡是参考他人成果的部分，要进行引用。
- 援引他人原话时，不得私自改动；应使用引号标注，写明来源。
- 转述时，必须用自己的话表达，不得照搬原文，确保做好引注。
- 切勿把他人的作品当作自己的作品。
- 决不能向不同的两门课提交相同的论文，除非预先获得双方教师同意。
- 不得买卖或"借用"他人论文，要独立完成写作。

下一章我将进一步说明如何诚实地写作，但核心思想不变：独立撰写，准确引用，标注出你引用的话或观点，公正地呈现所有资料。

可以上网购买论文吗？

坚决不要。

买卖论文是欺诈行为。具体来说，你把从朋友那里抄来的、网上下载的或用两三篇文章拼凑的论文提交上去，本质上都是欺骗。这跟朋友事先允许、网络信息自由获取毫无关系。把别人的成果说成是自己的，是欺骗。把自己的作品"借给"别人，同样是欺骗。

这类作弊经常暴露。老师们阅文无数，嗅觉敏锐，很容易寻到种种迹象，看穿不诚实的行为。有些文章一看就不是某个学生的风格，还有一些作品离题甚远，甚至前后矛盾。

---

❶ 引自亚伯拉罕·林肯《葛底斯堡演说》。
Abraham Lincoln, Gettysburg Address, November 19, 1863, reprinted in Gary Wills, *Lincoln at Gettysburg: The Words That Remade America* (New York: Simon and Schuster, 1992), app. I, pp. 202–3, and app. III D 2, p. 263.

比如一篇论文在开头写道"It's for sure America will sign to this international treaty.";到了收尾又得出结论说"Having ratified this pact with such lofty ideals and soaring hopes, America will soon confront its harsh realities..."。写出第一句蹩脚英文的学生是不大可能在后面一句华丽转型的,结尾定是出自他人之手——由此可见,学生未使用引号,剽窃他人语言。还有一些论文的问题出在参考文献上,例如文献失效或过于陈旧。时至今日,如果你还写"President Lincoln will probably defeat the Confederacy and win reelection",那么我也是倾向如此,我只希望林肯没去那家害他遇刺的剧院。

在甄别网络抄袭和剽窃方面,教师们有丰富的工具和经验。既然可以轻而易举地搬运网络信息,老师们见招拆招,将可疑的语句放入谷歌搜索,瞬间即可真相大白。此外,大学都与一些检测平台合作,对学术不端行为进行鉴别。检测平台汇集了庞大的数据库,包括公众可用的所有网站、几乎所有已发表的论文,以及历届学生提交的大量文章。教师可以借助这类平台,将学生新提交的论文和数据库进行比对,自动得出查重结果;平台吸纳这些新作,数据库不断扩大。

请别把整个过程当作一场猫鼠游戏,这其实关乎诚实的工作、诚实的人格和真正的学习。购买论文是欺诈行为,是对学术诚信的直接玷污,学校完全有理由严肃处理。对于学生而言,这种行为是在主动毁掉自己的前途。

2.4 使用网络资源

现如今,做学术、搞研究都离不开网络。它能带你瞬间浏览多个信息源、对比不同作者的侧重点、获取实验所需的特殊材料……网络能做的还不止于此。它既是优质的科研工具,也是人们掌握事实的捷径:美国女性何时获得选举权?印度人口比中国多吗?印地500系列赛(India-napolis 500)的赛道总长大约是多少英里(1英里=1.609千米)?

使用网络资源时,请记住三条注意事项。

第一,网络上信息的质量参差不齐。任何开通了网站、博客或推特账号的用户都能畅所欲言,信息质量很难得到保证。如果 *Encyclopaedia Britannica*(大英百科全书)网站显示:1944年6月6日是诺曼底登陆日,你可以放心采纳;但如果 *Encyclopedia BetaTest* 声称:1984年4月1日诺曼底登陆,你最好到别处查查。要选取可靠的信息源,还要通过不同的来源交叉核对。如果你不清楚哪里的信息更可靠,可请教老师或图书馆里的参考馆员,他们能为电子数据库的使用提供帮助。

有一部百科全书值得一提,那就是 Wikipedia(维基百科)——大家都在用,教师们也不例外。把它列入论文的参考文献时,要小心对待,反复核对。维基百科上的文本完全由志愿者撰写,他们的知识水平和见解都有很大差异,但你却无从知晓具体情况。志愿者贡献的信息也没有专门的编辑负责核实。所以我的建议是:可以使用维基百科,但请你慎重行事,尽量为重要信息提供额外的论据。引用维基百科时,要仿照其他信息源的处理方式,做好注释。

第二，网络检索的效率极高，无论身在何处，你都可以瞄准一个话题、深入挖掘信息，缺点在于：查询越有针对性，得出的检索结果反而不会包含背景信息，但若想理解某个议题并探究其价值，背景或语境往往至关重要。例如在搜索框内输入"ISIS + 2018"，你将得到数千条"伊斯兰国"极端组织发动袭击的消息，然而你可能不会了解到它与中东不稳定的政治局势或欧洲、加拿大、美国的国土安全有何关系。如果你需要上述信息，建议补充阅读，并尝试用"ISIS and Saudi Arabia"或"ISIS and homeland security"等关键词重新发起检索。

第三，把网上的信息复制到电脑实在是太轻松了，这对整理笔记极为有利，却也会造成很大的麻烦。既然网络资源在科研工作中不可或缺，我要列出一些潜在的问题，请你时时注意。首先，跟阅读、理解信息相比，复制信息（包括扫描和复印）要简单得多。其次，许多人会因此抵挡不住诱惑，一心想要作弊。只要你没有把复制来的部分在论文里标记清楚，就是剽窃。最后，即使你本不想剽窃，也很容易把复制来的部分和原创的部分混在一起，搞得自己也无法区分。

要如何应对这些问题呢？其一，要对信息来源进行筛选。一些来源比如许多的在线期刊，只发表专家评议过的高水平文章，并且拒收所有不达标准的文章。还有一些来源走向另一个极端，这些网站的运营者会发邮件给你，请你提供银行账户和信用卡信息，理由是他们的父亲，一位腐败又慷慨的百万富翁，愿意与你分享财富。

网络的入口无人把守，也没有编辑为信息质量提供保障，你只能自己擦亮双眼，判断信息源的可信度。

其二，为了真正理解某一议题，要判断是否需要掌握上下文语境和大的背景信息。如果需要，建议适当扩大探索范围，你可以上网检索其他关键词（上面的案例就用到了"ISIS + Saudi Arabia"或"ISIS + homeland security"），或沿着页面内的超级链接深入阅读有关背景。一些重要文献或有影响力的书籍也可视情况列入阅读范围，它们通常包含议题所处的政治、社会和历史背景。这跟射击有点类似——目标是正中靶心，但它周围的圆环同样不能忽略。

其三，整理笔记时，避免过多摘录网上的内容。你到后期很难分清到底是谁写了哪一段，实在是自讨苦吃。确有必要从网上摘录时，要合理区分这些内容、详细记录其来源，并且应该在笔记里自始至终采取统一的标记方式。信息的来源要具体到页码，先是写在笔记上，随后要录入论文里。等到你一边写论文、一边在笔记里写写画画时，你又需要把粘贴的话跟自己的话区别开。我将在第 3 章分享一种简单的方法给你，它对书本、文章和电子资源中的引文都有效——只要在引文的开头、结尾分别标注大写字母 Q 即可。大写字母 Q 在笔记中非常显眼，标记起来十分简便，手写的笔记也可采用这个方法。

如果摘录的内容来自网站，一定记得留下网址或 DOI 编码❶，便于引用和回访。鉴于

---

❶ 数字对象标识符（DOI, digital object identifier）是一串数字和字母的序列，用于永久标记一篇文章、一份文件，具有唯一性。即使改变了一份文件的网址，其 DOI 保持不变。

网上的信息变动频繁，建议你同时记下访问日期，以满足某些引用格式的要求。如果你使用了 Medline 或 MathSciNet 这样的专业数据库，要记下文章的 ID 号，这都是为了方便引用与回溯。

> **使用网络资源的建议：**
>
> - 参考网络信息时，要自主审查信息质量。
> - 适当扩大检索范围，到背景材料中查找关键的语境信息。不能只盯着检索结果里的只言片语，还要浏览相关的书目和文章。
> - 避免把过多的内容直接搬运到笔记中。要用自己的话对信息进行整合，你能从中学到更多。
> - 要有一套清晰的标记系统，单独区分复制粘贴的内容，比如我提到的字母"Q"标记法。
> - 记清网站的地址和读取时间，方便日后引用。想回顾对应内容时，也可节省时间。记录数据库内的文件 ID 也是出于这个原因。

2.5 语言课

大学里的外语课很常见。幸运的是，在学外语方面保持诚信并不难。规则直截了当，你一定不会感到陌生：作业和考试都要独立完成，不得从同学、解题手册或已发表的译文那里抄袭答案，不得使用手机辅助完成翻译。基本原则依然是：当你说你做了某项工作，你必须确实做了。汉语普通话培训课程是如此，中国艺术、中国历史也不例外[1]。

能不能使用汉英词典呢？每个老师的要求不一样，还要考虑作业任务的性质。一般来说，家庭作业允许查词典，在这个过程中你可以积累新词、重温旧词。到了学校，一些课内的测验或翻译练习不允许使用词典，目的是检测你对词汇的掌握程度。一位同学从阿拉伯语课上总结出一条重要经验，她说：尽管所有的任务都允许使用词典，但是随堂的翻译测试有严格的时间限制，花太多时间查字典反而没有好处[2]。在线词典尽管速度更快，恐怕也会来不及。不要把词典当成拐杖，不肯放手，用它来学习词汇、答疑解惑是更好的选择。

讲完了这些基本问题，你还需要知道保持诚信如何能够切实助力外语学习，这很重要。

一位意大利语老师是这么说的："我要把学生培养成好的听众、好的读者，并鼓励他们自发地进行口头和书面表达。"如果你在学习古典语言，你需要做到快速阅读、准确翻

---

[1] 此处的普通话、中国艺术、中国历史等课程指的是美国大学开设的相关课程。——译者注
[2] 一些学生出于合理原因确实需要更多的时间，例如学习障碍，或是听不懂授课语言。如果你需要更多时间完成考试或作业，要提前向老师说明情况。学校可能会要求你以书面形式提交申请。

译。但具体到掌握任何一门语言，唯有通过实践——也就是所谓的听、说、读、写、译，所以才会有家庭作业和各类练习，它们带你走上实践之路，营造沉浸式的语言环境，帮你循序渐进地了解一门语言。除了能练习使用语言，你还获得了向错误学习的机会。一位老师跟我说，习得语言的过程就是改正错误的过程，错误扮演了重要角色。

这就解释了为什么不做作业、照抄答案会极大削弱学习效果。很显然，每一位跟我交谈过的老师都深知"实践出真知"这个道理。在完成练习的同时，你能巩固旧知识，还能趁机发现并改正错误。不要觉得能把几段话翻译正确就大功告成了，理解一门语言远远不止于此，无论是西班牙语、俄语还是日语，你既要掌握一般的规则，还要知道如何运用特定的单词和语句。

已发表的译文或解题手册就在那里，不用岂不可惜？如果你的老师不介意，偶尔可以拿来参考，不过仅限于下面两个场景：复核答案或应对棘手问题。一位拉丁语教师告诫学生说："遇到翻译难题要先自己思考，尽力应对。试过之后才能对照现成的译本，看看自己处理得是否得当，或是学习译者的解决思路。"换句话说，他人的译文可以用来辅助自学，但不能代替自己的工作和思考。

物理、化学、数学以及经济学系的教师建议你用同样的方式看待各类解题手册。请它们来做无声的老师，帮忙检查你的功课。千万别用它们替代你的功课，欺骗他人、坑害自己。

适用于所有课程的建议：

解题手册、已发表的译文、他人的成果通通不能代替你自己的工作。如果老师允许，你可以偶尔将它们用于自学，比如核查功课或克服特定的障碍。

2.6 小组作业

课程中的大部分作业都是个人项目，还有一些需要组队完成。小组可由教师指定，也可自愿组合。一位科学家说："我告诉学生，小组任务是最高效的学习方式。大家聚在一起研究考试可能遇到的问题，结果收获了分析同一份材料的不同视角。组里的三四个同学会从不同的侧面解读某条信息，他们全程记下的笔记都是不一样的。"

加入小组的第一件事是搞清老师的要求。课程、任务不同，具体要求也随之变化，了解清楚才能诚实地工作。有些任务会要求小组在一起学习，但是以个人为单位提交书面作业；有些任务要求以小组为单位提交报告或现场陈述；还有一些只需要互动、讨论，不设置写作任务。你需要知道你所在的小组是哪种性质、要完成什么任务。如有疑问，及时与老师沟通。

小组作业的建议：

> 要明白老师有怎样的预期，如小组的任务是什么，个人的职责有哪些。在开始行动前，要把这些都搞清楚。如果小组里有人不做事，要先跟对方私下沟通，讲不通，再向老师求助。

如果要求是作为一个小组来学习，然后分别提交个人论文，那你就不能把其他同学的成果直接用在自己的作品里。你们可以彼此交流、互相学习，但是不能共享书面作业，不能照抄同伴的观点。一位老师告诉我，这种做法一眼就会被看穿。

另一种情况是严格意义上的团队任务，比如向全班做小组展示。这需要每个人为此付出，而且不能浑水摸鱼。以小组为单位展示成果时，要列出每个人的角色和贡献。在协作中体会团队的价值，本身就是一种收获。一位科学家说："小组任务是学生为未来的工作、生活进行预演，我们每个人取得的成就大都离不开与他人合作。"只不过，团队里确实会有人企图蒙混过关，与他们相处也是一种考验。如果有人作弊或者拒绝合作，可以先找对方私下协调；必要时，请老师介入。

以上是关于小组作业的建议。当然，无论是个人项目还是团队项目，只要任务派发到个人，你就得独立做好分内的事。与他人交换想法是正常的，你可以跟队友或朋友讨论作业的内容，见面、邮件、短信或是网络聊天室等多种方式任选。但你们不能互换答案，或者直接把别人的答案提交上去，这是作弊行为。你需要自己组织语言，自己做完分配到的任务。

既然合作是允许的，那又该如何把握分寸，避免"过度合作"呢？不同的科目或者老师对此有不同的界定。几乎所有老师都会支持"讨论"这一行为，他们反对的是以寻求帮助为名，把任务分包给别人（团队有分工的情况除外）。举例来说，你要完成的个人项目是编写电脑程序，那么写代码、编辑代码以及调试软件，这些都是你的职责。你需要保证产品的可用性，排除出现的故障。你必须不借助任何外力做好一切，否则，整个项目就失去意义了。但是我得承认，其中有一定的灰色地带，比如请人来检查程序的语法，有的老师会鼓励学生互相帮个小忙，当然也会有老师禁止。不确定的时候要主动咨询，这也是我能给出的最好的建议了。计算机程序设计、宏观经济学、细胞生物学或许都在你的课表上，到底哪些工作需要自己做、哪些可以结伴做，任课老师最清楚不过。

为学习小组布置习题任务

最常见的小组作业就是合作解题。数学、经济学、统计学或者说理科类的专业普遍钟爱这个项目。老师们这样安排自有一番道理，合作解题是一种很棒的学习方式。

然而，为了避免作弊的嫌疑，还是需要明白老师为"合作解题"制定了哪些规则。适合某位老师或某套题目的规则，在另一处未必适用。有些情况是小组共同提交一份答案，有些时候却对此明令禁止。规则随着作业的变化而调整，就像开卷、闭卷考试一样，没有对错之分。拿到每套习题，都要听清对应的规则；如果感到疑惑，可以要求老师解释，然后按照规则执行即可。

有两件事值得注意。第一，避免学术不端。如果你的名字单独出现在论文上，意味着是你自己单独解决了问题。考虑到小组的存在，你自然可以咨询同伴的建议，但是所有书面工作须由你独立承担。另一方面，如果你以合著者的身份出现在团队论文上，你必须承担了相应的工作。

署名的建议：

如果论文上你是唯一作者，说明全部工作由你一人完成。如果你是合著者之一，你必须为团队项目做出了应有的贡献。

第二，如何最大限度地通过小组作业提升自己。下面谈到的行为其实算不上作弊（或者说处于模糊地带），却能影响你的教育体验。如果小组由四人构成而且习题又刚好能切分为四份，你是否赞成每人各领一份？虽然一些老师同意学生把任务拆解开来，但是并非最好的学习方式。毕竟到了考试的时候，你得一个人搞定所有题目，不能只回答其中的四分之一，也不能把难题转移给队友。

实际上，"得到答案"是不够的。从别处抄一抄最容易了，但这无异于自我毁灭。学习的目的不仅在于得出答案，更在于找到通往答案的最佳路径。和终点相比，旅途同样重要。教师布置的习题提供了验证路径的机会，没准儿还能让你意识到自己已经迷路了。在练手的过程中及早发现问题，这很值得庆幸，真到了"生死攸关"的考场上就太迟了。

因此我建议组员们各自为战时，要兼顾任务中的所有题目，最好全都做一遍，然后再到团队里集中处理部分（或所有）习题，互相检查、寻找修改思路，共同进步。这样一来，你对所有材料了如指掌，遇到论文和考试也能应对自如。

完成习题任务的建议：

即使是合作解题的项目，也建议每个人先独立完成所有题目，再以小组为单位讨论、核对。这不是"学术诚信"的必要条件，但可以帮你更好地学习。

看不懂一些题目的解法该怎么办？不确定哪种方法是对的又该怎么办？可以麻烦队友帮忙讲解吗？答案是肯定的，你甚至可以向小组以外的同学求助。但前提是，你必须已经尽力试过了，如果障碍依旧存在，就跟朋友、室友、老师、导师、助教以及小组成员多聊聊吧。注意：请教的焦点是解题方法，看能否诊断出问题所在，答案的数值反而是次要的。如何检验学习成果呢？你得能说出为什么这种解法和技巧行得通，别的方式却不行。

学习外语、数学、科技和经济学的时候都可以这么做。一位老师的描述更为具体，他说："当初我教拉丁语入门，布置了一些难题让学生讨论，结果发现同伴交流对解题很有用。不过，把题目全推给别人是不行的，一定要自己先挣扎、先思考。"这位老师还建议：若是别人的帮助起到决定性作用，你得在书面作业中如实记录，并致以感谢。这并非强制性要求，但不失为一种值得推荐的做法，注明一切以后，当事人和读者都能分清楚谁做了什么；教师也能精准捕捉到一些普遍存在的难点，适当增加指导。

确保团队学习效果的建议：

在合适的环境里遇到对的人，学习效果将大大提高；在喧闹的环境里遇到错的人，纯粹是浪费时间。组建团队之后如何让学习更加有效呢？一些经验丰富的本科生跟我分享了如下建议：

- 队友尽量不要超过三个。团队越大，聊天的时间越久。闲聊倒是没什么，但会扰乱正常的学习安排。
- 碰头的地方需要安静、可交谈并且不用担心打扰到其他人。图书馆会开辟一些专门的空间给学习小组使用；学校周边的咖啡馆也不错，挑一些顾客少的时段碰面。宿舍里分散注意力的东西太多了，不推荐把讨论安排在那里。(移动设备最容易干扰注意力，请关闭手机。)
- 为小组学习设置明确的起止时间。团队应达成共识，按计划准时开始每次讨论。讨论的时长一般定为 60~90 分钟，为了应对期末的重大项目可适当延长。预先定好结束时间能督促团队提高效率。
- 碰面前，要选定讨论主题，确保全员知情并各自做好准备。不要对主题进行拆分，让组员分别负责不同的部分，这么做看似高效，实际上又往往不尽人意。身为组长，如果手上有提纲或指导手册，要提前圈定讨论的要点；如果团队的任务是解出 10 道题目，要保持所有人进度一致——通知里不能只说"讨论分两次进行"，而是应明确每次要讨论的题号，保证大家提前做完相同的题目。
- 建议选择符合下列特征的同学组建团队：
  ○ 具有跟你相似的学术水平。这样更容易步调一致，而且彼此都有输出，而不是单方面的持续输出。
  ○ 努力上进，并且愿意为团队付出。避免跟不学习却想"躺赢"的同学组队。

再强调一次，团队项目提倡合作精神，需要的是一个双向互帮互学的环境。
如何找到对的人呢？某位同学提到了一个非常棒的方法："我会观察身边的同学，看哪些人每次都到课，而且从不迟到。我格外关注那些言之有物的同学，他们的发言不是为了讨好老师。我相信他们在团队里也会非常乐于分享。"

除非明确规定了允许多人署名、共同提交一份作业，组内学习时，你不可越界操作，直接把别人的答案抄下来交给老师。"别人的答案"也包括解题手册或旧习题集里的内容。凡是以个人名义提交的作业，都必须自己组织语言、自己作答。尽情讨论、积极贡献，向组内组外的伙伴学习，不必为多次请教他人感到自责；但是接下来，亲自把事情做完。这便是诚实的学习。

跟同伴一起上台做题

几个人站到黑板前一起做题，为相互抄袭提供了极大便利。只要台上有一个人先把物理（或者工程）题目解好，教室里的所有人都能迅速得到答案。一位来自数学系的课程负责人对我说，"大多数老师认为这是作弊，而且雷同的作业看起来非常明显。"

先别急着辩解说看黑板是出于无奈，这位课程负责人想介绍一个更好、更诚实的做法给你：把题目都解出来，单独提问或集体讲授均可，确认每个人都懂了；然后把板书和答案擦掉，让每个人自己再答一次。这个环节绝不是无用功，实践和学习到这一刻才真的开始。

万一在黑板上写答案的是助教怎么办？助教为全班服务，这样抄下来的答案就不是作弊了。唯一的问题是：照抄一遍不代表学会了，因此刚才的建议依然作数：为了弄懂解题过程、掌握技巧，就不能满足于机械地复制答案。万一助教略过了关键的步骤，或你对解题策略仍有疑惑，一定要求助、询问。

关于小组作业最后想说的话

一位科学家是这么定位团队任务的："总体说来，我让学生讨论是希望他们一起找到解决问题的方法。但具体的操作要由单个人执行，这就是所谓的'在做中学'；只有亲自练习了、体验了，才有可能获得相应的技能。"要想验证你有没有获得这些技能，可以举办一场答辩，面对组员回答或是自问自答：为什么这个就是正确答案？你如何一步一步得出答案的？

老师们还会不厌其烦地强调积极参与小组活动的重要性。一位物理学家认为，"消极被动没有任何好处。从队友那儿拿到答案从来都不是首要目的，你得领悟解题过程，学习应付难题的策略。"但前提是你得参与其中。第一步，跟团队一起做功课，把握解题过程；第二步，自己动手，把题目做一遍。遇到瓶颈，及时寻求帮助。然后自己再做一次。

把材料理解通顺以后，自己动笔写出答案，避免复制现成的语句。即使老师同意你们联合创作，上述做法也是最佳选择。单纯把答案抄一遍不等于理解了，更不幸的是，这些"不理解"导致的学习差距不会立刻显现，而是会在考试、论文或者更高深的课程中一下子暴露出来。

> **如何最大限度地通过小组作业提升自己，建议如下：**
>
> ● 在小组中要表现得积极主动，不能消极被动。
> ● 关注解题方法和思考问题的方式，而不是答案的数值。当你真正掌握解题过程之后，自然能解答作业、论文、考试中的问题，还能捍卫自己的答案，为别人解释其中的原理。

从这个角度出发，公开出版的解题手册阻碍了学习进程。这些手册极易在现实或网络中买到，它们为主流教材的习题提供答案。一些老师是同意使用这类参考书的，另外一些老师则反对，但无论如何都不建议过度依赖解题手册。的确可以把它们视为学习资源，用来快速核对答案、提供反馈。可如果有了手册就不做功课，自己的学习将极大缩水。回想小时候学骑自行车，后轮两侧的稳定轮不能永远辅助我们，最终总是要取下来的。

解题手册还会威胁到学术诚信。如果你的答案是从书上抄来的，那就是作弊。同样性质的行为还有抄袭其他同学的答案或者从网络下载答案，都是拿别人的成果冒充自己的。

诚实地完成小组作业的建议：

有些课程会把学生分组，但作业还是每个人单独提交，此时你就要清楚哪些事可以团队做、哪些事必须自己做。提醒如下：

- 提交作业前，务必理解老师的具体意图：什么事应该由小组一道来做，什么事应该由学生独立做？
- 放心去跟小组成员研究习题、讨论课内话题，向他们请教不懂的方法或答案。你还可以反复求助，直到自己完全理解。
- 但是，如果别人直接把答案提供给你，无论是写在纸上、黑板上还是来自现成的解题手册，都是作弊；如果你把答案抄下来，假装是自己写的作业交上去，也是作弊。对你的学习而言，这就是灾难，在未来的考试和更高阶段的工作中，你将自食其果。

2.7 加入实验室

除了习题任务，另一种常见的团队项目就是做实验。你要跟实验室的同事一起调试仪器、进行实验、收集数据、应对爆炸危险，甚至要一起呼叫消防部门。最理想的是，你们始终遵守安全操作规程。一起做好这一切以后，你们要分头完成实验报告，各自得出结论。

实验室里的规则非常直接——按任务要求，集体做实验，但需要独立做好完整的实验记录、独立撰写实验报告。没有老师的明确许可，绝不能借用他人数据或作业。

准确的实验记录

每次实验，都要精确记录你的实验步骤和结果。不同实验室有不同的记录方式，有些使用传统的实验日志，有些使用计算机程序或软件，还有一些使用针对特定实验设计的问卷。在任何方式下，都要尽快完成结果录入。别以为自己能把数值背下来——人的记忆并不可靠。不要篡改结果数据。若未得到教师允许，不得采用他人数据。

事先学习你所在的大学和实验室有关实验记录的规定，了解其严格程度。不知如何采取行动时，及时询问老师或实验室助理。在科研方面享有盛誉的莱斯大学为此编写了详细的注意事项，学校认为：严谨、全面的实验记录对维护研究的有效性以及可复制性有重大意义。

A well-kept notebook provides a reliable reference for writing up materials and methods and results for a study. It is a legally valid record that preserves your rights or those of an employer or academic investigator to your discoveries. A comprehensive notebook permits you or another researcher to reproduce any part of a methodology completely and accurately.❶

老师们深知其中的道理，也都是用同样的记录流程要求自己的。如果你感到困惑，可以向他们请教。

与实验室同事一起工作的建议：

齐心协力，完成实验。但是每个人都要各自做好实验记录，并独立提炼研究结果。

在一些特殊情况下，老师会允许参考外部数据。比如老师可能会让实验彻底失败的学生根据他人数据完成分析；还有可能是老师用一些数据向全班举例。加入实验室就是为了学习如何做实验、如何处理数据，偶尔借助他人数据也是出于这个目的，但是这种做法必须征得老师同意。如果不确定，及时询问。

实验记录的建议：

● 提前与实验教师或实验室负责人沟通，了解实验记录的要求，例如是否需要用到实验日志、电脑，或为特定实验编制的一系列表格。
● 每结束一个实验，尽快录入数据。不要依赖自己的记忆。
● 如实记录你在实验中犯的错。之后，用正确的方式完成实验，并在日志中工整地划去出错的部分（如果使用了软件，跟随程序提示操作）。如果数据记录在纸质日志上，不要把出错的那页撕掉，也不要用涂改液遮盖，要确保原始数据清晰可读。如果数据已录入电脑，不要删除，可以将旧数据备注为"incorrect"（错误），或为文本添加删除线。在老师的帮助下，你能更好地吸取教训，从错误中领会新知识。

---

❶ 引自莱斯大学《实验记录细则》，引用日期：2017 年 11 月 18 日。
Rice University, "Guidelines for Keeping a Laboratory Record," Experimental Biosciences, accessed November 18, 2017, http://www.ruf.rice.edu/~bioslabs/tools/notebook/notebook.html.

科学家们的传统做法是把实验结果记录在专用的笔记本上，这种笔记本每一页都标有页码并且分册装订，想要在上面动手脚、掩盖错误是很困难的。"专用"就体现在这一点上。一位化学老师提醒："绝不要在这种笔记本上使用涂改液，也绝不要撕掉其中的页面，因为页码都是连续的。如果实验出错，把相关内容工整地划掉，在后面继续记录即可。"

为何要保留这些错误呢？一位科学家的回复是："在实验中，与结果同等重要的是你的观察数据和你的操作流程。万一实验没成功，根据你精确记录的步骤和结果，老师们可以帮你找出问题。"实验记录代表着一切工作"正在进行"，为人们实时展示你的每一步操作。这样的记录才算达到数学或物理老师的要求。对他们来说，你的每一步操作都反映出你思考问题的方式。遇到困难时，追踪这些记录就能找到问题出在哪个环节，不至于搞错方向。

目前，大部分实验室都用电脑取代了传统的手写笔记，并且配备了数据分析软件。技术手段日新月异，但数据的录入规则保持不变：

- 做完实验，尽快录入。
- 精确记录实验结果，失败的实验也不例外。
- 重做某个实验时，不要抹去旧数据，可将其标记为"incorrect"。保留完整的数据有益于反思、排错。还有一种可能是到了后期，你突然发现旧数据根本没错。
- 若无明确指令，你始终都只能使用自己的数据。

做好这些，下一步你就要独立得出研究结论了。

诚实的数据

诚实做实验的本质是如实陈述实验数据。所有纸质、电子的实验记录应该真实、完整、可靠，它们准确地呈现你的每一步操作和你得到的实验结果，哪怕实际数据没有达到预期效果。

切记不可捏造实验结果，不可借用他人实验结果，也不可因个人需求调整结果。禁止随意修改结果。不得本末倒置，先画出完美曲线，再创造数据迎合。

美国科学基金会将此类现象列为"科研不端行为"，所谓的"不端"是指"开展科研活动时，在准备、获取、报告研究结果过程中发生捏造、篡改或抄袭等严重偏离公认做法的行为"[1]。不管你的研究有没有得到美国科学基金会的资助，始终谨记：不要在数据上作弊。

只是，作弊的诱惑一直都在。比如，由于你的拖延，许多事来不及了，偷偷编几个数据或从朋友那"借"点儿数据就能迅速结束战斗；又如，实验结果跟预先想好的不一致，要是回过头彻查原因呢？测量结果有无偏差，设备调试有无疏漏，有没有可能出现了低级

---

[1] *Federal Register* 56 (May 14, 1991): 22286–90. Additional rules on research misconduct are in *Federal Register* 65 (December 6, 2000): 76260–76264.

错误……还是把"漂亮"的数字直接填进去最痛快。要知道,即使已经把标准流程读得滚瓜烂熟了,每个人在实验时还是会出现差错,忠实地记下你做的一切,然后去跟正确的步骤不断对比,这会加深你对科研的认识。

一位生物学家指出,最常见的实验错误发生在实验开始之前——阅读操作指南的时候不够仔细,或是忘记把所有要用到的设备和材料提前安排好。实验进行到中途,如果少放了某种化学品或少用了一个烧杯,再想补救,为时晚矣。井然有序的实验能让你学到更多,你会得到更可靠的数据来开展下一步工作。

> **建议:**
> 开始任何实验之前,仔细阅读操作指南,准备好所有你需要的仪器和材料。同时,做好记录数据的准备。

实验结束后,老师或实验室主管可能会查看你的实验记录。这是求之不得的好事。作为课程的初学者,你能跟他们学到最好的实验方法;如果是更高阶段的项目,他们能帮你评估研究进展、增进协作,并确保整个实验室都在诚实地工作。

实验室的学术诚信取决于诚实的数据,诚实的数据出自实验室的每个人,无论学生还是教师。任何层面的个人或集体对任何类型数据的违规操作,都将得到严肃处理。日本一位杰出的细胞生物学家被迫撤回多篇文章,原因是他就职的大学认定他在五份出版物中伪造了图像[1]。跟多数骗局的曝光过程类似,其他科学家无法复现其研究结果,由此引发了对他的调查。杜克大学因涉嫌骗取超过 2 亿美元联邦政府拨款而遭到起诉,学校在 60 份拨款申请中使用了虚假的生物数据[2]。一位物理学家通过造假,声称发现了 116 号和 118 号元素,被伯克利的劳伦斯利弗莫尔国家实验室开除[3]。这些案例的涉事者都是资深科学家,但是同样的标准也适用于做实验的学生。每个人都必须按规则行事,科学研究的基石才不会崩塌。

不要误以为只要科学家完整、诚实地呈现数据就够了,每个领域的研究者都该如此,同时还要确保参考文献真实可靠。曾有一本关于美国早期枪支所有权的获奖图书,它引经

---

[1] Aggie Mika, "Investigation Finds Cell Biologist Guilty of Misconduct," *Scientist*, August 2, 2017, https://www.the-scientist.com/?articles.view/articleNo/50015/title/Investigation-Finds-Cell-Biologist-Guilty-of- Misconduct/.

[2] 在这起科研不端行为的标志性案件中,被指控的科学家认罪。她丢掉了工作,并不得不撤回多篇已发表的论文。如果举报人诉讼成功,杜克大学将面临高达 6 亿美元的处罚。Jef Akst, "Duke Sued for Millions over Fraudulent Data," *Scientist*, September 6, 2016, https://www.the-scientist.com/?articles.view/articleNo/46974/title/Duke-Sued-for-Millions-over-Fraudulent-Data/. "Duke Admits Faked Data 'Potentially Affected' Grant Applications," *Retraction Watch* (blog), June 29, 2017, http://retractionwatch.com/2017/06/29/ duke-admits-faked-data-potentially-affected-grant-applications/.

[3] Robert L. Park, "The Lost Innocence of Physics," *Times Education Supplement* (London), July 24, 2002.

据典,却引发质疑,因为没人找得到作者参考的诸多关键文件❶。外部专家介入调查,证实作者存在严重的学术不端行为,该教授随即离职。

在做到数据诚实的基础上,还要在呈现它们时做到公正合理。故意隐瞒坏消息是不可取的。假如你正在验证一种假设,而某些实验结果跟你的假设不匹配,你完全可以核查整个实验,找寻可能的错误,然后再试一次。所有这些操作都应该准确体现在你的实验记录中。一旦你修正了问题、完成了对假设的再次验证,无论结果正面与否,你都要完整、坦诚地展示实验结果。

诚实做实验的建议:

● 跟实验室的同事一起做实验、一起讨论,但每个人都要有自己的实验记录,还要独立整理实验结果。
● 千万不要抄袭或伪造实验数据。
● 不要省略或隐藏那些你不希望出现的实验结果。所有结果都应真实、完整地存入实验记录中。
● 诚实地公布你的实验结果,即使你认为它们是"错的",即使它们与你的假设相抵触。

每个学科都要遵守这些规则,历史学、社会学、政治学、经济学等学科也不例外。刚刚提到的那本有关枪支所有权的著作,有确凿证据表明它对历史数据的呈现存在"严重失实"。外部专家发现,该作者采取回避策略——某些反对他观点的论述,书中只字未提。

对任何科目而言,隐瞒不利的结果都有违基本的学术伦理。在你的论文或演讲中只报告积极结果,这很诱人,但你不应该这样做。请记住,如果是法庭案件或辩论比赛,你可以有选择性地呈现事实,以帮助你的一方,但学术研究不同,你应该诚实、完整地介绍结果,包括所有的缺陷。这才能确保科研工作不断向前发展。基于完整的数据,你可能会想出更好的假设或更有力的解释,别的研究者也同样有机会加入进来。在任何情况下,你都应该完整、准确地呈现你的结果。

---

❶ Michael A. Bellesiles, *Arming America: The Origins of a National Gun Culture* (New York: Alfred A. Knopf, 2000). 贝莱西尔曾是埃默里大学历史系教授,并兼任该校暴力研究中心主任。

❷ 由于找不到贝莱西尔引用的某些关键文献,批评者们纷纷抗议。埃默里大学任命了一个由知名历史学家组成的外部委员会调查此事。委员会得出一系列证据确凿的结论,并使用了"严重失实"的说法。他们说,"根据贝莱西尔教授列出的时间或地点,没人可以得到'枪支拥有率低'这一相同结论。"调查还发现,贝莱西尔排除了与他的结论相矛盾的数据,特别是 Alice Hanson Jones 有关更高枪支拥有率的证据。委员会最终对贝莱西尔的学术诚信"提出严重质疑"。Stanley N. Katz, Hanna H. Gray, and Laurel Thatcher Ulrich, "Report of the Investigative Committee in the Matter of Professor Michael Bellesiles," Emory University, Atlanta, July 10, 2002, http://www.emory.edu/news/Releases/Final_Report.pdf; Michael Bellesiles, "Statement of Michael Bellesiles on Emory University's Inquiry into Arming America," Emory University, Atlanta, July 10, 2002, http://www.emory.edu/news/ Releases/B_statement.pdf.

2.8 课堂参与

许多课程，特别是研讨课，会将你在课堂的参与程度列入总评成绩考核。老师们喜闻乐见的场景是：学生积极投入讨论，彼此倾听，提出有价值的问题，并根据阅读材料形成有深度的观点。这是提升教师获得感的最佳方式了，对学生又何尝不是呢。

在课堂讨论环节，"学术诚信"应该不是大问题，困扰一些学生的是如何界定"参与"以及如何为此赋予分数。既然课堂参与度常常是老师的主观判断，学生有疑问也就在所难免。还有一个原因是老师们极少针对"参与"列出详细的规定，学生不太清楚评价的标准。

尽管我们无法完全排除评分中的主观因素，对于"有效参与课堂"还是比较容易描述的：

- 你正常去上课了吗？
- 你是否在课前做好准备，完成了老师布置的任务，并且所有的发言和提问都有充分的论据支撑？
- 你的参与和贡献是否有效推动了课堂讨论的进程？

贡献的方式多种多样，可以是提问、作答，还可以是评价、延伸他人的发言。但为了确保高质量的参与，你需要按时完成任务，虚心听取老师和同伴的发言。这样的话，同伴之间、学生跟阅读材料之间才可以形成有效互动，通过想法的碰撞，每个人都能有所斩获。

研讨课跟纯粹的讲座不一样。听讲座的时候，学生安静就座，动笔记录，偶尔提出一两个问题。研讨课（或是理论课每周都有的讨论环节）则是基于课程任务，有明确方向性的对话。如果你提前不做准备、到场一言不发，或者单向输出、讲话声盖过所有人，你的收获也会非常有限。一位研讨课的负责人告诉我："我们不鼓励被动的学习，我们希望学生在学习中发挥主动性。"

要做到积极主动，就要提前完成阅读任务。阅读不能走马观花，目标是充分理解材料，辩证地看待其内容。从初次阅读直至随堂讨论都要带着这样的心态去做。

在课堂上，不要时刻想着把自己知道的一切都展示给老师，把它们留到考试和论文里。也不要想着跟所有人达成一致意见。只要你保持开放的心态，尊重他人观点，持论有据，从独特的视角阐述观点也是没问题的。就像一位老师所指出的那样，"对话和辩论是高等教育体系的基本准则。"

即使你尚无十足把握，你仍然可以发表新观点、揭示新角度，就当是为你的创意进行试水。你的任课老师以及我本人都期待你在小组讨论中抛出这些想法。不过，你得事先做足准备，确保完成了阅读任务和家庭作业。

因此，研讨课最需要的是参与对话、探究教学材料、同学间坦诚地交换意见，从而形成更活跃的学习氛围，带来更丰富的学习体验。整个互动的过程也能显示出你看待问题的方式以及你对问题的思考。

如何参与课堂讨论并从中受益，建议如下：

- 按时上课，不迟到。
- 高质量完成派发给你的任务，不拖延。
- 认真听取老师和同伴的发言。
- 通过提问、回答、回应他人的方式来推进讨论。

最后补充一条建议：争取在初始的一两节课就参与到讨论中，哪怕是简短地说上一两句话都可以。这是为了让你尽早打破沉默——不要安静地坐在一边，等那个让你夸夸其谈的完美时刻。课堂讨论而已，不必刻意追求完美的发言。课堂讨论倡导的是在交互中学习，既是交互，就要有取有予，有问有答，也就意味着有错误和失误。相信我，人人都会犯错。如果你在前几次课选择默不作声，你很容易一直到最后都犹豫不决，不敢参与。最简单的解决方案就是：主动在早期加入讨论，即使是只提一个小问题或只点评一句话。一旦有了这样的经历，再参与到对话里就顺畅多了。

老师们在意的恰好也是这件事：让讨论持续进行，让尽可能多的学生参与其中。一位很有经验的老师为了鼓励发言，可能会先邀请几名学生点评阅读材料或回应前面同伴的观点。如果你被选中，别觉得老师在找你麻烦，他这么做不是想考验或恐吓学生，而是希望有越来越多学生自信地跟全班分享想法。

如果你比较羞涩或者不习惯在课内发言，要试着走出舒适区。如果临场发挥让你感到焦虑，不妨提前写下阅读材料中的一两个观点，在课上发言时针对它们提出问题或者发表评论即可。如果在小组里发言让你感觉不自在，可以在办公时间找到老师，尝试在那里开启一段交谈。体验过这种一对一的交流，你可能会更安心地参与课内讨论。你还可以到学校的学业指导中心寻求帮助，咨询学习技巧，获取学习建议。

课堂参与的建议：

争取在前两次上课时说点儿什么，一些简短的评论就可以。它会给你未来更多的课堂参与打开一扇门。

另一个建议：在研讨课和各种讨论环节，尽量到前排就座。懒洋洋地缩在后排，躲在所有人后面，极易导致消极被动地听讲，不利于主动参与讨论。

2.9 低分申诉

确实有那么几次，你一拿到老师发回的实验报告、论文或试卷，就觉得分数太低了。听完答案和解析，你更觉得自己回答得足够高明，唯一的问题在于老师给分不够慷慨。对此，我的建议是：先给自己一点时间冷静一下，仔细想想，再复核一遍自己的答案。很有可能你在回看的一瞬间，发现方才自己眼中的高手只是错觉。

如果还是觉得自己值得更高的分数，当然可以跟老师沟通，提出低分申诉。但是你自己必须事先复查试卷，想好申诉理由。"我真的需要一个更好的成绩啊！"——称不上是理由。你要合理地解释为什么你的答案比它目前的分数更好。

保持礼貌谦逊也会有帮助。尽量让谈话向着积极、有建设性的方向发展。咆哮着表达不满或愤怒地质疑评分者的能力，到头来只会激怒老师。一般来说，你不应该使用任何包含"蠢""笨"等字样的话语。为了合理地提出诉求，你可以说："我发现我的回答跟您上节课给到的思路很相似，可以请您再看一遍我的答案吗？"或者说："您说过回答这种问题必须涵盖三个要点，我觉得我在答卷上都涉及了。"还有就是："要想取得进步，我该怎么做？"——在几乎所有时刻，这句话都推荐你使用。

成绩申诉、复查有标准的流程，你需要了解清楚。比如是否要以书面形式陈述诉求，正式提起申请；在一些班级规模比较大的课程中，向教师申诉之前要先由助教核实、协调。

如果你的理由充分，也许能如愿以偿，拿到更高的评分。即使分数没有改动，你也可以知道自己哪里做错了，下一次尽量做得更好。不要对一时的分数耿耿于怀，将来的课程里，你还有许多"下一次"。

不少学生常犯的错误是缺少完整的解题步骤，尤其在数学、统计学以及其他的理工类学科中。只写出最终结果是不够的，即使求解正确，也不能得满分，因为你没有呈现全部的过程。把你做的所有工作包括数据和流程都提交上去，这样既能排除作弊嫌疑，又能让评分更加合理。老师可以为回答正确的部分赋以相应分数，如果求得的结果错误，老师还能找出问题所在，引导你妥善修正。

展示你做的所有工作，就相当于展示你思考问题的方式，这才是老师想要看到的。正如一位物理学家所言："光写下答案，几乎不具有任何价值。我们做教育，就是想让学生学会解决问题：如何提出问题，如何完成演算，如何为得到的结果提供解释。也可以说，我们要看到学生如何运用物理学思维……如果一个学生不给我看到这些东西，我就没法做出评判。"

所以从小学三年级开始，每位数学老师都在反复唠叨：要写过程，不能只写答案！这绝对是一条忠告。

最后，如果你为成绩提出申诉，必须做到：在重新评分之前，不以任何方式修改你的答卷，否则就是作弊。

低分申诉的建议：

- 在决定是否申诉之前，再次复核自己的答案。
- 礼貌地解释申诉的具体理由。
- 在重新评分之前，不在答卷上做任何改动。

2.10 荣誉行为守则

一些学校用荣誉行为守则来确保学术诚信。学生们通常会签署一份承诺书，保证诚实地开展工作，并相互监督，报告违规行为。他们不仅要为自己的诚信负责，还必须报告其他人的作弊行为。同时，学生承诺在课外认真行事，对自己的行为负责，并在评判违规行为时发挥主要作用。

违反荣誉准则的问题通常由荣誉顾问委员会而不是学院院长来处理。一些委员会完全由学生管理，当然也有一些委员会吸纳教师和行政人员加入。无论哪种情况，学生在维护荣誉准则运行的方面都发挥主体作用——学生为学术诚信承担责任，用实际行动突出学术诚信在教育中的核心地位。

至于诚实工作、负责任行为的实质内容，学校之间并无太大差异。学生们许诺：独立完成任务；不抄袭、不作弊、不购买论文；遵守老师们关于论文、实验报告和考试的规定。在课外，绝不骚扰、恐吓或威胁他人。大多数大学都推行此类规则，无论这些学校是否订立了荣誉行为守则。

区别在于，荣誉准则需要学生们为了维护并促进这些高尚的道德标准主动承担起个人和集体责任。一名学生是这样概括的："同学们不断地评估自己在课堂内外的行为。不只是同学之间的日常互动、学术诚信，还有学校理事会为学生团体分配资金的方式等，所有行为都算在内。"

老师们依靠学生自行遵守诚信制度，同学们则相互信任、彼此制约。例如，课内考试不设置教工监考，或者允许学生在图书馆、宿舍进行闭卷考试。老师们相信学生能坚守荣誉准则，而且每个人都知道，大范围的违规将摧毁整个教育体系。

荣誉准则的确意味着抓出作弊行为，也意味着监督、惩罚等一系列应对措施，但是其目的无疑不在此。正如一位同学所说："监视使人们感到被低估和不被信任，这与荣誉准则的意义背道而驰。荣誉准则的存在，是为了让学生安心做出自己的道德选择；它还让学生知道，自己被当作成年人一样平等对待，周围的人会用理性、同情和理解的态度与自己相处。"

相似的话，我一次又一次听到学生提起。他们说，建立荣誉制度是为了一个积极的目标，它最大的价值是在学术和社会生活中塑造诚实和负责任的风气，让正直、诚信成为一种强大共识，直到每个人都自觉遵守。

相比于规模庞大的研究型大学，荣誉行为守则在教学型学院中更为多见。这并不奇怪。在教学型学院模式下，同学们彼此了解，有强烈的社区意识，有利于这些守则发挥作用。荣誉行为守则也确实构成了这些群体的核心精神支柱，甚至成为学院、学校自我定位的关键。这些守则鼓励公平和诚信的学生文化，唤起个人和集体的责任感，还能增进学生和教师之间的信任。

不管你的学校是否采用荣誉制度，上述目标都是值得向往和追求的。

3

优秀笔记养成秘籍

好的笔记可以帮助你从阅读材料、研讨课和讲座中学到更多的知识。无论你是在准备期末考试,还是在写研究论文,你都需要整理材料,梳理主要观点,并记下自己读过和听过的内容,此时笔记就要派上用场。在这一章中,我将告诉你如何有效地记录笔记并避免常见的问题。

最重要的方法,其实也是最简单的——上课认真听讲,为所有的讲座、阅读材料整理笔记时,都经过认真思考。抄录笔记不能代替听课和阅读本身,当然也就更没必要把你听到、看到的一切都转化成文字。笔记不必面面俱到,笔记旨在留存真正重要的东西。

一位生物学家告诉我,"单单是做笔记这个简单的行为就对听课有帮助。它迫使学生集中注意力,摘录讲解中有价值的部分,并预测接下来会出现的内容。"同理,你在读书、读文献或学习其他资料时,也应该做笔记。笔记能帮你更清晰地理顺手头的资料,加深理解。事后回顾笔记,还可温故知新。

> **建议:**
>
> 阅读、听讲时都要做笔记,还要复习你的笔记。记得为记录的内容提炼主题,把它写在页面顶部,例如"Coleridge and English Romantic Poetry"。如果是讲座笔记,要标明课程、章节、日期等信息;如果是书面材料,还要把详细的著作信息记录在案。

3.1 何为优秀的笔记

既然要做,就努力做到最好。但是什么样的笔记才称得上优秀、有效呢?在撰写论文、复习考试或预习新课等场景中,都能充分发挥作用,就是优秀笔记。具体来说,你的笔记需要:

- 突出讲座或阅读材料的要点及结论。
- 阐明要点之间是如何联系的。
- 提示你用到了哪些论证方式和证据。

你还要把所有的定义,以及最重要的等式、公式和运算法则都备份到笔记里。当然在英语课上,你可能听不到太多的等式。

同时,记录一些有用的细节也是有必要的:经过筛选的优质引文、关键的日期,或是论文和考试中用得到的数据。总之,你得评估一下,如果内容本身很重要,或者能用来说明更宏大的问题,就可以把它们包括到笔记里。但是不能让你的笔记被各种细枝末节所淹没。笔记是为理解重点内容服务的,重点内容是指论点、论证方式和论据,以及把这一切有机串联在一起的逻辑主线。这些是最值得你关注的内容。别把自己当成录音软件,你不需要捕捉、保存对方的每一个字。

建议:

笔记的目的不是记录一切,而是要突出要点,并把它们合理地组织到一起。

做笔记要有选择性,这一点得到了研究的支持。研究表明,手写笔记比键盘录入更有利于学习。原因也显而易见。电脑可以很方便地复制粘贴语句,无需操作者进行太多的思考。手写就不一样了,学生得对内容做出取舍,记录的过程相当于思考的过程,这就促进了对信息的深度加工,到考试时也能发挥得更好。

在我看来,这不是让你立刻把笔记本电脑换成纸质的螺旋笔记本,无论笔记记在哪里,你都要分清主次,然后还要认真、多次复习。把内容抄写下来不等于学习。

为了炼成优秀的笔记,你需要:
- 在讲座和课堂讨论中,认真听讲。
- 选择真正重要的内容,记在笔记中。
- 课后和考前都要复习笔记。
- 复习笔记时,继续在旁边标注新的想法。

优秀的笔记能清晰、简明地呈现要点,是你复习功课的好帮手。不过,单纯地浏览笔记是不够的。要想彻底理解课程材料,除了重复,你还得多思考、多质疑,搞清所学内容的来龙去脉。高质量的笔记是一个好的开始,它表明你在主动地学习,而不是机械地抄录。

---

[1] Pam A. Mueller and Daniel M. Oppenheimer, "The Pen Is Mightier than the Keyboard: Advantages of Longhand over Laptop Note Taking," *Psychological Science* 25, no. 6 (April 23, 2014): 1159–68, https://doi.org/10.1177/0956797614524581.

3.2 听课笔记和读书笔记

听课笔记和读书笔记有许多区别，一些比较明显，一些不太容易察觉。阅读书籍、文章和其他资料的时候，你能自己把握阅读和做笔记的节奏。听课的时候，你得一路跟随老师的思路，所以最初的笔记通常不完整，需要后期补全。尽量趁记忆深刻，在上课的当天就把笔记整理好。原则不变，注重要点和有用的例证，不要关注次要细节。教育的关键任务，就是让你学会辨别真正重要的事物。请用优秀的笔记展示出你的辨别力。

一般说来，为文章、书籍和网页整理笔记要比为讲座、研讨课整理笔记容易得多。这是因为书面材料以完整的状态示人，已经形成清晰的内在结构。作为读者，我们可以看到一篇文本的不同部分、每个部分的标题，因此可以借用这些信息来搭建笔记的框架。以罗伯特·利伯所著的 *Retreat and Its Consequences*（《撤退及其后果》）为例❶，其中一章讨论了美国对欧洲的政策，作者分析了"The Cold War Era and global order""What's wrong with Europe?""Europe and a reluctant USA"等问题，那么你在整理笔记的时候，就可以使用对应的小标题。

讲座也是有结构的，只是相比之下不是特别清晰。若想能抓住核心内容，你需要格外关注两个部分：引言和结论。同时，注意抓取讲座中的信号词，例如 the three main theories are 或 turning to a slightly different topic，这样你就能判断出内容的逻辑走向，然后在笔记里为三条主要理论留出位置，或者另起一行写下你听到的新主题。

> **建议：**
>
> 要为笔记划分出不同的部分，以此来体现讲座或阅读材料的结构。

谈完了节奏和结构，来看笔记的内容。听课笔记往往是匆忙潦草、前言不搭后语的；读书笔记却经常写得废话连篇，详细至极。尤其是看到网上现成的文本，很容易产生搬运的冲动——这种原封不动的粘贴，一定要控制数量。你必须时刻提醒自己：复制不等于阅读，你仍然要专门花时间把材料读完，然后选取特定的部分粘贴存档。千万不要对所有内容一视同仁，抄完一段又一段，这种做法很有诱惑力，但是缺少了去粗取精的环节，形成的笔记也就不够实用。别忘了，笔记的内容要合理划分为不同的部分，分类管理，突出重点。

---

❶ Robert J. Lieber, *Retreat and Its Consequences: American Foreign Policy and the Problem of World Order* (Cambridge: Cambridge University Press, 2016).

建议：

优秀的笔记总能让真正重要的内容突显出来。如果你用电脑做笔记，可以为文本设置加粗字体。如果是手写笔记，可以借助箭头、下划线来强调重点内容。如果对记下的某个部分存疑，建议标记三个问号（？？？）来提示自己，表示需要进一步查证或找机会向老师、助教、同学请教。

对笔记做整合的时候，务必要区分他人的成果和自己的原创。否则，当你用这些笔记来撰写论文时，很可能无意中把复制来的文字当成自己的观点写入正文。这是非常严重的错误。即便你不是故意为之，但要向老师或系主任解释，打消他们的疑虑，就没那么简单了。

幸好这类错误可以避免。我将在后面的一节介绍几个笔记技巧，其中就包括字母"Q"标记法，用来解决内容混淆的问题。这样，你就能准确分辨他人的话语和自己的评论，也就可以放心大胆地参考自己的笔记了。

建议：

为书面材料整理笔记时，要把他人的原话标注清楚，还要注明具体页码。这是为了防止使用引文时无意识的抄袭行为。建议你在引文的前、后分别加上大写字母Q，使其在笔记中清晰可见。

上面分析的是听课笔记和读书笔记的区别。二者的相似点也值得留意。无论何种场景，笔记的功能都是帮你回忆要点，把它们连贯地串在一起，使之形成有意义的整体。阅读、听讲的时候要多注意引言和结论，笔记也应该充分呼应这两个部分。如果有机会，把引言、结论多读几遍。引言对主要议题进行预告，结论对要点进行回顾和升华，这些恰好都是你需要的内容。

建议：

细读某本书或某篇论文之前，把引言、结论和各部分的标题浏览一遍。这么做可以快速统揽全局，让阅读更有目的性和方向性，还能提高笔记的质量。

3.3 复习和改进你的笔记

优秀的笔记是复习和深度思考的基础。不能做完笔记就把它丢在一边，这样之前的努力都浪费了。要最大化利用自己的笔记，多翻阅、常复习。我的建议是：写完笔记的几个小时内就及时回看一次，考试或写论文前还要再看一次。

第一次回看有两个目的。其一，回忆相关的材料。其二，对笔记进行注解和扩展，尤其是听课笔记。由于是一边听讲一边速记，你有时会发现笔记里有过于简短或过于晦涩的部分，趁此机会，抓紧完善必要的内容。

复习的同时，还能再度改进自己的笔记。可以把老师列举的例证填进去。可以把用词模糊的地方澄清一下。还可以穿插自己的观察和得出的结论（我的做法是把自己的评论写在括号里，就像现在这样）。此外，你还可以为某些部分添加标题。比如，你回顾了有关法国大革命的听课笔记，希望增加"The Fall of the Bastille"和"Deposing the King"两个小标题，整个讲座的脉络就更直观了，据此来复习备考也会更有效率。

> **建议：**
>
> 听完讲座或读完文章最初整理的笔记还不算完成。你仍需抽出一点时间做两项重要工作：第一项是复习笔记。第二项是增补注释、扩展完善。电子版的笔记编辑起来更方便。如果是手写的版本，补充内容可填写在页面空白处或笔记的结尾。

两项工作都不会耗费太长时间，几分钟就能做完，带来的好处却不少。首先是加深印象。其次，你能进一步从宏观上把握课程内容。最重要的是，这两项工作把你从信息的被动接收者转化为主动的学习者。

每次复习笔记，都尝试回答下列问题：这里主要在讲什么？能说出其中最关键的三到四个要点吗？老师或写作者是如何层层推进的？提供了哪些证据？内在逻辑是什么？哪个例子最恰当？然后，你应该能够用自己的话复述这场讲座或整篇文章的核心要素。如果你无法用简洁、平实的语言把它们说清楚，你就没有真正理解。但幸运的是问题出现在复习环节，一切还来得及。尽快把材料再读一遍，或者请教老师或助教。

3.4 用字母"Q"标记法做笔记

一些诚实的写作者被指抄袭，并因此陷入巨大的麻烦，罪魁祸首便是糟糕的笔记体系。时间久了，这些作者分不清哪些是自己写的，哪些是先前抄来的。为了避免此类问题，你要在笔记里做好区分。

最好能找到一种简单的方法，把抄录别人的语句跟自己的原创观点、文字分隔开来。

常见的解决方案是使用双引号，然而实际的效果并不理想。原因是：第一，引号太小，不够明显。为了写论文从笔记里选取参考资料时，极易无视引号的存在。第二，引号不包含引文的页码信息，无法满足引用格式的要求。第三，如果引文内部含有另一层引文，把各种引号梳理清楚就比较有挑战性了。

我想提供一个更好的方案：在笔记里引入大写字母 Q 就能免除各种混乱了。在引文的开头写上字母 Q 以及对应的页码，在引文的结尾再次以字母 Q 结束。做起来很轻松，而且字母 Q 在笔记中非常醒目，写论文或复习的时候一眼就能看到。

每当你开始为一份新的材料整理笔记，第一件事就是在笔记的开头记好作者、标题等基本信息（具体涉及哪些信息，我将在本书的第二部分详细介绍）。每一本书、每篇文章，以及网上的资料都要如此，这是引用格式的需要。有了这些出版信息，再加上字母"Q"标记法，你在引用时就可以完全信赖自己的笔记，不必跑回去频繁翻查原始文档。

使用字母"Q"标记法突显引文的建议：

Q157 Churchill's eloquence rallied the nation during the worst days of the war. Q

这种标记法的特点是简单、清晰、有效，手写笔记和电子版笔记都可使用，还能解决引号套引号的问题。回看笔记时，你可以准确定位作者的原话以及它们所在的页码。你还能知道作者是否引用了其他人的著作。最后，笔记上所有字母 Q 范围以外的话语，都是你自己转述的内容。

转述的建议：

你转述的版本不能太接近作者原话。如果不确定，仔细对比两个版本的措辞。

字母"Q"标记法还能处理复杂的引文，我们分不同的情况来看。第一种情况，引文语句在原始文献中跨页出现。要想标出分页的位置，可以在引文中插入双斜线（//）。它跟字母 Q 一样，易于辨认。这依然是为了准确地引用——你可能在写论文的时候决定只截取引文的一部分，有了双斜线标记，就能引用准确的页码，无需再去核查原文。请看下例：

Q324–25 Mark Twain's most important works deal with his boyhood on the river. He remembered // that distant time with great affection. He returned to it again and again for inspiration. Q

由此可知，引文的第一句话位于第 324 页，引文的第二句话跨过页面，引文的第三句话全部位于第 325 页。以上信息全部由字母"Q"标记法和双斜线提供，而且你任何时候查看笔记都一目了然。

另外的一些复杂情况是，你或许希望删去某些不必要的部分，或是为了理解顺畅增加少量词语。针对这两种情况，其实都有明确的规则。

3.5 使用英文省略号缩短引文

对于直接引用，我们的要求是准确无误，但是如果按照下面两条规则操作，你可以将引文适当缩减。第一，删减过后的引文仍需保持原意。第二，你必须用省略号准确向读者指示出省略发生的位置。英文的省略号是三个点（...），省略号的前、后要有空格。

如果你在引文某一句话的中间省略了一些词，直接插入省略号即可。

【原文】　　　　　　I walked downtown, which took at least thirty minutes, and saw her.

【缩减后的引文】　　I walked downtown ... and saw her.

如果引用的内容来自两句话，裁剪之后剩余的两个部分可以合理地拼接为一句话，在引文中间也应该使用省略号。

【原文】　　　　　　I walked downtown. After walking more than thirty minutes, I rounded the corner and saw her.

【缩减后的引文】　　I walked downtown ... and saw her.

鉴于省略号的出现有时会打乱句子的原貌，缩减过的版本不妨再检查一遍。请记住：省略号纯粹是为了表示你有意在引文中省去了某些部分。省略发生的位置决定了是否要在省略号基础上使用句号。

表3.1　省略发生的位置和标示方式

省略发生的位置	省略的标示方式
在一句话的中间； 或在两句话之间，但缩减后的引文以一句话呈现	只使用省略号即可
在两个分句之间，但缩减后的引文以不同的句子呈现	保留原文断句处的句号，再紧跟一个省略号（省略号后面的引文以大写开头，即使这节引文原本不在句首）

只要不改变原作者要表达的意思，并且按上述规则用省略号做好标记（表 3.1），就可以将引文缩短。

3.6　使用方括号为引文增加可读性

由于引文脱离了原文语境，你偶尔可能想在里面增加一两个词，让引文的语义更加明晰。比如，你援引的那节原文里出现了人称代词，而你希望将指代还原，让读者知道这些人的名字，那么在维持原意不变的基础上，你要释放信号给读者：自己在一些地方对原文做了微调。方法是以方括号为标志，把增补的词放进方括号内。请看案例：

【原文】　　Q237 Secretary of Defense James Mattis was speaking in New York that day. The President called and asked him to return to Washington immediately. Q

假设你只打算引用第二句话。对读者而言，他们看不出总统召见的对象是谁，原封不动的引用反而表意不明。稳妥起见，你需要把"他"的身份补入引文，并在新增的文字两边使用方括号。所以引用的版本写成：

"The President called and asked [his secretary of defense, James Mattis,] to return to Washington immediately."

尽管你借助方括号对原文做了调整，这句引文仍然是准确无误的。如果上面的版本只修改了文字，却忘记加方括号，那就是引用错误。

再提醒一次，对引文所做的增补或删减均不可改变原文含义，并且要按规则使用方括号、省略号。"引文"意味着这些引用的内容另有所属，你无权变更甚至歪曲原作者的意思，但欢迎你在论文中表达赞同、反对，或提出自己的解释。

3.7　引文内部含有另一层引文

你引用的那节原文里可能自带引号，这些引号与字母"Q"标记法互不冲突。也就是说，引文内部的引号无需特殊处理。例如：Q47 He yelled, "Come here, quick," and I ran over. Q 或许你还记得上文提到过的常见笔记方案，与之相比，字母"Q"取代了笔记中的第一层引号，就不必担心引号套引号会导致误解了。

3.8　字母"Q"标记法的综合应用

到现在为止，我已经描述了字母"Q"标记法、省略号、方括号的基本用法，还有引文套引文的应对方法。这个笔记体系能用来记录十分复杂的引文，并完美对接你的论文写作过程。我们来看一个综合应用的案例：

Q157–58 Some of Churchill's most famous speeches // were actually recorded by professional actors imitating his distinctive voice and cadence. ... The recordings were so good that [one friend] said, "I knew Winston well and still can't tell who is speaking." Q

由这段笔记可知：

- 引文最前面几个词出现在第 157 页，其余的部分位于第 158 页。
- 原文中"cadence"后面的一些内容被省略了。
- "cadence"后面先是句号，然后是省略号，说明引文的第一句话以"cadence"结束，后面才发生了省略。
- 方括号内的"one friend"不在原文中。
- 引文的最后出现了对别人的引用，原作者把它放入引号中，表明那句话是别人的原话。

掌握了这套笔记体系，未来写论文的时候，你就可以直接从笔记中引用整理好的内容，无需翻找原文，而且不会发生无意识的剽窃。解释整套体系好像花了很长的时间，但我可以保证它用起来是省时省力的。

4

剽窃与学术诚信

前一章介绍了做笔记的基本方法,目的是诚实、有效地完成论文写作。这一章我们来展开细节,谈一谈如何避免各类问题。

最大的问题是剽窃,即把别人的成果说成是自己的,这些成果涉及文字、证据、数据、图纸、代码或想法等。剽窃行为是对学术规则的严重破坏。一旦被发现,后果很严重。事实上,被发现的概率极高——肇事者的论文或课程都可能被评为不及格,甚至会因此导致停学或开除。这可不是一张违停罚单那么简单,而是高速公路车祸,倘若肇事者是有意为之,则是不系安全带的高速路恶性车祸。

剽窃的情况不多,但确实偶尔发生,有时候还是无意间犯错。如果读书笔记做得杂乱无章,无法区分作者的文字和自己的点评,等到用笔记写论文时,就会不经意地把本属于原作者的话当成自己的原创。即便这是无心之失,面对多疑的老师或院长,你可能也会百口莫辩。

好消息是,在笔记中引入字母"Q"标记法就能理清一切,同时预防上述问题。具体做法请回顾第 3 章的相关部分。

当然,糟糕的笔记并非造成剽窃的唯一原因。赶着提交论文的学生有可能忘记做好必要的引注。一些学生粗心大意,一些学生不清楚引用规则。更不幸的是,还有一些学生故意作弊❶。

不管理由为何,无论肇事者是本科生、研究生还是教师,剽窃都是一种严重违反学术规则的行为。把别人的话语或想法错误地说成是你自己的,构成欺诈。请记住学术诚信的基本原则:当你宣称自己完成了某项工作,你实际上确实做了。如果参考了他人成果,需要如实引注。直接引用时,要明确标注、写清出处、勿做修改。呈现研究资料时,要做到公正合理、真实客观。不得篡改、歪曲或伪造引文、数据、实验以及他人的观点。

---

❶ 引用未加标注是错误行为。最令人不安的情况是,故意使用另一作者的成果而不注明出处,这就是剽窃的经典定义。有些人扩展了这一概念,认为剽窃还包括无意间的复制和挪用,我将此称为"无意识的剽窃"。即使是无心的借用——糟糕的笔记所致,而非蓄意偷盗——它仍然是严重的问题。无论你是否愿意称其为剽窃,它都是对学术规范破坏力极大的一种行为。

4.1 合理引用他人成果

引用他人成果以公开、诚实为基本原则。如果使用了他人原话，必须清楚地标明是引文（可以用引号或缩进），同时将引文的来源列入参考文献。仅仅提一下原作者的名字是不够的。如果是直接引用，请使用引号，并提供完整的出处。如果是转述别人的话，要换成你自己的语言表达，不要大量套用作者原话，然后记得合理添加参考文献。

使用了他人的视觉图像、建筑图纸、数据库、图表、统计表、计算机算法、口头报告以及网上的信息，应遵守同样的规则。参考他人的成果，要进行引用。哪怕你认为这件作品是错误的，打算写文章批评它，也要做好引用。即便该作品能在公共领域免费获取，也要引用。即使作者允许你使用该作品，依然要引用。所有这些规则都出于同一个考虑：坦白承认你从别人那里得到了什么。唯一的例外是：当你陈述众所周知的信息时（至于什么是"众所周知的"，取决于你的目标读者或听众）。例如谈到万有引力，你不需要给艾萨克·牛顿进行专门的注释。

不遵守这些规则会受到严厉惩戒。对于学生，后果也许是挂科，甚至开除。对于教师，也许会降级或者失去教职。这一系列严厉措施都说明学术诚信是大学的根本。

避免剽窃的建议：

对于论文中涉及他人的贡献和成果，要给予充分认可。
- 如果你使用了其他作者的原话，把它们放入引号，并合理引注。
- 如果你转述了其他作者的话，要用自己的语言表述，不得模仿原文；同时要合理引注。
- 如果你参考或提及了其他作者的观点，无论你想表达对它的赞同或反对，都应注明其来源。

4.2 合理引用网络资源

使用网络资源的时候要格外小心，务必遵守相应的引用规范。如果不需要阅读长篇大论的文章，上网查找资料是非常高效的，这是网络的优势所在。你可以做大量有针对性的检索，快速查看多个信息源，访问优质的数据库，点击选中文章的摘要或关键语句，然后将素材复制到你的笔记中。这些都很好。实际上，这算得上是开展研究的最佳方式，当然也是最常见的方式。然而，把所有信息分门别类地梳理清楚，也是至关重要的。你需要一种简单、始终如一的方法，把其他作者的原话和你转述的内容妥善记录下来。

最便捷的方案是坚持使用第 3 章介绍的读书笔记体系：为粘贴到笔记里的所有内

容增加字母"Q"标记（引文的前后都要添加）。在此基础上，你可以改变这部分的字体类型，还可将字体颜色改为红色或蓝色。只要在笔记内部保证格式统一即可。这样一来，即便是过去三四个星期，等你回顾笔记、从中选取材料撰写论文时也肯定不会搞混。

还有一件事：一定要把网站的网址或 DOI 复制到你的笔记中，方便日后的引用和回溯。建议你将访问的日期一并记下，某些引用格式需要附上这一信息。如果文档出自某个数据库，且带有 ID 号，也请准确地录入笔记。

4.3 合理地引用和转述

这一小节通过一个案例来说明如何合理地引用和转述他人话语（表 4.1），以及如何避免常见的错误。下面的表格列出了引用和学术诚信的一些主要规则，我们借用 Jay Scrivener 写的一段关于 Joe Blow 的描述，数字 99 代表对应文献的编号。

表 4.1　合理地引用和转述

【原文】Joe Blow was a happy man, who often walked down the road whistling and singing.	引自 Jay Scrivener 所著的 *Joe Blow: His Life and Times*
正确做法	
"Joe Blow was a happy man, who often walked down the road whistling and singing."[99]	正确：完整的引文放在引号内，后面紧接着指向引文的出处 *Joe Blow: His Life and Times*
According to Scrivener, Blow "often walked down the road whistling and singing."[99]	正确：照搬原文的部分都在引号内，标注了参考文献。引用的部分没有误导性
"Joe Blow was a happy man," writes Scrivener.[99]	正确：对原文的部分引用，照搬原文的部分都在引号内，标注了参考文献
According to Scrivener, Blow was "a happy man," who often showed it by singing tunes to himself.[99]	正确：对原文的部分引用，为 whistling and singing 做的转述放在引号外，转述内容准确传达原作者的意思，而没有照抄原句。句后标注了参考文献
Joe Blow seemed like "a happy man," the kind who enjoyed "whistling and singing."[99]	正确：两处部分引用都在引号内，非引用部分在引号外。标注了参考文献
Joe appeared happy and enjoyed whistling and singing to himself.[99]	正确：合格的转述。没有套用 Scrivener 的措辞。标注了参考文献
错误做法	
Joe Blow was a happy man, who often walked down the road whistling and singing.	错误：这是剽窃，原因是与原文措辞相近或直接引用了作者原话，未加引号，未做引注。一定要注明引文的来源
Joe Blow was a happy man, who often walked down the road whistling and singing.[99]	错误：未加引号，却引用了 Scrivener 的原话，这依然是剽窃，即使添加了准确的文献信息。只要是直接引用，就要使用引号（或为较长的引文增加缩进）。因此，左侧的句子需要给出明确的信号，告知读者这句引文其实是作者原话

Joe Blow was a happy man and often walked down the road singing and whistling.	错误：尽管这不是作者原话，但与之非常相似（只是变换了 whistling 和 singing 的位置）。可以改为直接引用，或转写成措辞明显不同的版本。无论采用哪个方案，都应标注参考文献
Joe Blow was a happy man.	错误：共有两处错误。第一，照搬作者原文的部分需要添加引号和参考文献。第二，左侧的句子看起来像是有别于原文的一个版本，但是应该标注引用，因为它属于原作者 Scrivener 的个人判断：Scrivener 认为 Joe Blow 是快乐的（而不是一个单纯的事实）
Joe Blow often walked down the road whistling and singing.	错误：与上一句的两个问题类似。第一，作者的原话须加引号，并标明引用。第二，这是 Scrivener 的个人判断，不要让读者误以为这是你的判断
Joe Blow appeared to be "a happy man" and often walked down the road whistling and singing.[99]	错误：虽然标注了引用，但是仍有 Scrivener 的原话在引号外面。这会让读者误认为引号之外的内容并不属于 Scrivener，而是你的原创。仅这一处错误不构成太大的问题，类似于超速行驶了一小段路。但是如果这样的错误多次出现，且涉及大量的文字盗用，就是严重的剽窃事件了
"Joe Blow was an anxious man, who often ran down the road."[99]	错误：引文不准确。根据 Scrivener 的原话，Joe Blow 不是"焦虑"，而是"快乐"；不是"跑"，而是"走"。这段错误的引用不算剽窃，但也是有问题的。你应该正确引用，而且你的研究和写作应该是值得信赖的。如果这样的错误一再出现，会严重误导读者，或者出现最坏的情况，即人们觉得你是有意的，最终可能被视为学术欺诈（剽窃是另一种形式的欺诈）
Joe Blow "walked down the road" quietly.[99]	错误：引号内的部分引用是准确的，可是引号以外的话歪曲了 Scrivener 的原意。这也不算剽窃，但违反了学术诚信的基本原则，没有公正、准确地呈现材料。如果此类错误反复发生，或不断累积形成某种偏见（比如让 Joe Blow 成为一个厌恶音乐或偏好安静的人），就又构成学术欺诈了。说到底，这种错误是对读者的误导

上面的表格集中体现了单句引用的处理方法，实际写作的时候还涉及整段话或整个小节的文献引用。我们举例来说，假设你撰文讨论城市贫困的问题，其中一节着重围绕 William Julius Wilson 的有关论述展开。你自然可以直接引用也可以转述他的著作，但是需要在该小节列出你参考的那几条文献，以反映他的研究对你论文的重要性。你还可以在章节的开头，加入一段补充说明，例如在注释中写上"My analysis in this section draws heavily on Kathryn Sikkink's work, particularly *Evidence for Hope: Making Human Rights Work in the 21st Century* （Princeton, NJ: Princeton University Press, 2017），55–93."，或者直接在正文里加入一段类似的陈述。总而言之，你要公开承认其他作品对你的影响。直接引用的部分还要提供准确的引文出处。

4.4 如何转述

转述原作者语句时，千万不要过于贴近原文，那属于剽窃。不要说自己是一时疏忽，也不要以为加上了参考文献就万事大吉。

究竟怎样才能做好转述呢？建议先把作者的原文放在一边，思考打算传递出去的要点是什么。用自己的话将这个意思表达出来（标注参考文献），并与原作者的版本进行比较。如果能找出数个相同的用词，或是你的版本仅基于原句做了同义词替换，那么需要重新转写。重写时，尽量抛开原作者特有的语言和节奏。这有时候很难做到，因为原文在你的头脑中挥之不去，而且怎么看都觉得是最恰到好处的版本。不过，你还是要尝试，让你转述的句子和段落跟原文有明显差异，读上去、看起来都不同。

如果你不擅长用自己的话去转述某个想法，可以先整理思路，把主要观点写成一条简短的提示。接下来缓一缓，经过一段时间之后再开始转写。仍然是不看原句，只参考那条提示完成写作。写好后，与原句作对比，再把自己的版本调整几次就能写成了。记得替换用词，从不同的角度遣词造句。如果还是不满意，不妨放弃转述，采取直接引用（完全引用或部分引用均可）。也就是说，引文既能以直接引用出现，也能以你转述的形式出现，但绝不能把引文伪装成自己的作品。

为什么直接引用不是首选方案呢？在特定情况下，直接引用的确是最佳方案，写作者希望原汁原味地呈现原作者引人入胜的语言，或借助原文揭示原作者的重要信息。当富兰克林·罗斯福谈到珍珠港事件时，他告诉美国人："Yesterday, December 7, 1941—a date which will live in infamy—the United States was suddenly and deliberately attacked …"❶我不觉得有人会想要转写这句话，直接引用是最佳选择，原句有其历史意义。此外，分析小说和诗歌类作品时，也会大量援引作者原话，解读特定表达所蕴含的创造力。再者，文如其人，原文的某些用词能将说话者的个性、背景表现得淋漓尽致。

可见，直接引用应该适度，避免过犹不及。过量的直接引用反而会削弱其魅力。那些普通的语句、普通的想法，就不必再抄录原文，把它们转述出来即可。基本规则不要忘：要注明引文的来源，转写时不可亦步亦趋。

以上原则适用于整个学术界。大学新生应尽快了解、掌握，教师也需要严格遵守。美国海军学院的资深教授 Brian VanDeMark 就因为触犯规则而被解除原有职位。该教授曾出版多部广受好评的作品，但他 2003 年所著的 *Pandora's Keepers: Nine Men and the Atomic Bomb* 被指与他人著作有太多重合之处❷。尽管大部分可疑的段落都能看到注释，但经过前面的学习你应该知道，标注参考文献并不能彻底解决问题❸。

Robert Norris 将 Brian VanDeMark 书中有争议的段落汇编到一起，表 4.2 是少量的节选（Norris 甚至还搜集了一份更长的清单，指控 VanDeMark 涉嫌抄袭他 2002 年出版的 *Racing for the Bomb* 一书）。

---

❶ President Franklin D. Roosevelt, Joint Address to Congress Requesting a Declaration of War against Japan, December 8, 1941, http://docs.fdrlibrary.marist.edu/tmirhdee.html.

❷ Brian VanDeMark, *Pandora's Keepers: Nine Men and the Atomic Bomb* (Boston: Little, Brown, 2003).

❸ Jacques Steinberg, "U.S. Naval Academy Demotes Professor over Copied Work," *New York Times* (national ed.), October 29, 2003, A23.

表4.2 Robert Norris 汇编的 Brian VanDeMark 书中有争议的段落

Brian VanDeMark 的作品 *Pandora's Keepers: Nine men and the Atomic Bomb* (2003)	Richard Rhodes 的作品 *The Making of the Atomic Bomb* (1986) 和 *Dark Sun* (1995)
"... Vannevar Bush. A fit man of fifty-two who looked uncannily like a beardless Uncle Sam, Bush was a shrewd Yankee ..." (60)	"Vannevar Bush made a similar choice that spring. The sharp-eyed Yankee engineer, who looked like a beardless Uncle Sam, had left his MIT vice presidency ..." (*The Making of the Atomic Bomb*, 336)
"Oppenheimer wondered aloud if the dead at Hiroshima and Nagasaki were not luckier than the survivors, whose exposure to radiation would have painful and lasting effects." (194–95)	"Lawrence found Oppenheimer weary, guilty and depressed, wondering if the dead at Hiroshima and Nagasaki were not luckier than the survivors, whose exposure to the bombs would have lifetime effects." (*Dark Sun*, 203)
"To toughen him up and round him out, Oppenheimer's parents had one of his teachers, Herbert Smith, take him out West during the summer before he entered Harvard College." (82)	"To round off Robert's convalescence and toughen him up, his father arranged for a favorite English teacher at Ethical Culture, a warm, supportive Harvard graduate named Herbert Smith, to take him out West for the summer." (*The Making of the Atomic Bomb*, 120–21)
"For the next three months, both sides marshaled their forces. At Strauss's request, the FBI tapping of Oppenheimer's home and office phones continued. The FBI also followed the physicist whenever he left Princeton." (259)	"For the next three months, both sides marshaled their forces. The FBI tapped Oppenheimer's home and office phones at Strauss's specific request and followed the physicist whenever he left Princeton." (*Dark Sun*, 539)

来源：Robert Norris, "Parallels with Richard Rhodes's Books [referring to Brian VanDeMark's *Pandora's Keepers*]," History News Network, http://hnn.us/articles/1485.html（引用日期：2004年6月22日）。为方便起见，我调换了原表格最后两行的顺序，但文字未做丝毫改动。

遗憾的是，VanDeMark 在上述段落都没有标记直接引用，也没把 Rhodes 列入参考文献。以表格最后一行为代表的好几段话，简直是在照抄对方的作品——这类状况即使只发生一次，就足够拉响警报了。另外还有几处，跟其他作者的版本太过接近，惹人怀疑，然而关键是书中实在有太多这种问题了❶。总结 VanDeMark 拙劣的转述、不标注文献来源行为的表格还有好几个，它们来自不同的作者，每个人都觉得自己受到了侵犯。按照美国海军学院院长的说法，"这部作品成书时，记录和展示文献来源的全过程存在缺陷"❷。这位院长和 VanDeMark 本人将问题归结为工作上的麻痹大意，而不是故意剽窃（这也是为

---

❶ VanDeMark 不止从 Richard Rhodes 和 Robert Norris 的书中摘抄了词句，他还涉嫌整段抄袭 Greg Herken、William Lanouette 以及 Mary Palevsky 的作品，而且未加引注，未注明完整出处。一些章节处理得比较模糊，让人很难界定是否是故意剽窃；但的确能找到和他人著作几乎一模一样的段落，还有一些文字与他人的相似度极高，整体看来难以取信于读者。
关于 VanDeMark 的作品与其他书籍之间的平行对照，网络上也有类似的表格。可以访问 History News Network，"Brian VanDeMark: Accused of Plagiarism," http://hnn.us/articles/1477.html（引用日期：2003年5月31日），这个网页上包含多个对比 VanDeMark 和不同作者措辞的表格链接。

❷ Nelson Hernández, "Scholar's Tenure Pulled for Plagiarism: Acts Not Deliberate, Naval Academy Says," *Washington Post*, October 29, 2003, B06, http://www.washingtonpost.com/wp-dyn/articles/A32551-2003Oct28.html.

什么 VanDeMark 只是被降职而不是直接解雇）。不过，惩罚还是很严厉的，这表明大学的各个层面对剽窃现象都相当重视。

4.5 剽窃他人的观点

除了窃取别人的语句，剽窃还包括盗用别人的观点和想法。假设有一篇比较《麦田里的守望者》和《哈姆雷特》的论文，你读后很震撼❶。文章的结论是：尽管表层结构千差万别，但两部作品本质上是同一主题的不同变体——通过不安的内心独白，将一个年轻人的极度苦闷和精神骚动展现出来。如果你准备在论文里采纳这一有趣的解读，而且是用自己的话诠释了同样的理念，你仍需引用该想法的原始提出者。否则，你的论文将错误地暗示，你就是这个观点的首创者。如此一来，霍尔顿·考尔菲德会说你是个骗子。这段假设告诉我们：不要将他人的观点据为己有，借鉴他人想法的时候要合理引注。如果这个观点已经成为众所周知的常识，那可以另当别论，这也是唯一例外的情况。

4.6 歪曲他人的观点

在这两章，一个反复出现的议题是：你应该忠实地引用、展示他人的话语和观点，不要歪曲事实。转述时，不应改变作者原意，哪怕你并不同意该作者的看法。缩减较长的引文时，你必须标明省略出现的位置，并确保核心观点完好无损。

这里其实有两个目的。一个是在你自己的作品中保持诚信。另一个是为不同想法的充分碰撞营造公平的竞争环境。无论你是否赞成这些想法，把它们公正、客观地呈现出来是直面它们的第一步。这既是为了实现公平，也是为了提升你的文章质量。请用一个艰难、公平的测试来证明自己的想法更胜一筹，而不是在对比各种想法的时候暗中布局，诱骗读者认可你的优势。

尤其要避免的危险做法是，故意设立一组不堪一击的稻草人，然后你就能不费吹灰之力将它们驳倒。这不仅是欺诈，还是智力上的懒惰。相信我，如果你忠实地表述那些最强有力的反方论点，直面它们的挑战，阐明己方论点的价值，那么你的立场会更可信、更有效。

4.7 结论：转述和引用的正确方式

转述和引用规则的制定是以下列核心理念为基础的：

---

❶ 对比《哈姆雷特》和《麦田里的守望者》主人公霍尔顿·考尔菲德，这个假设是我自己想出来的，不过我猜测已经有人做过了。为保险起见，我决定打开谷歌搜索一番，结果最上面一条就在给我推销一篇相同主题的学期论文。沮丧之余，我又键入"Catcher in the Rye + phony"，这一轮搜索关联到的论文足以将我淹没。试想一下，如果有人买了一篇论述"霍尔顿·考尔菲德对骗子深恶痛绝"的文章来欺诈读者，那该有多讽刺。

- 你应该为自己的著作负责，也就意味着要对它所包含的观点、事实、分析与解释负责。
- 如无特殊标注，你作品中的每个字都默认是你的原创。
- 当你依据他人的成果或想法进行写作时，要公开予以承认：
  - 使用他人观点或数据，应提供文献来源。
  - 援引他人原话，应使用引号，同时指明出处。
  - 转述他人作品，应按照自己的语言风格转写，并做好引注。转述的版本不得模仿原文。若转写后仍与原文无异，改为直接引用。
- 当你动用前人的研究发现来完成论述时，要公正地展示这些成果。不得歪曲篡改。不可把错误的论点强加于对方，偷换概念伪造出一个稻草人，再将其击倒。
- 呈现实证研究的资料时，应说明是在何处取得，以方便他人自行核查验证（众所周知的材料无需额外进行引注）。

这些公平、公开的原则既作用于各类引用格式，又能确保你自己文章的各个部分井井有条。更重要的是，它们构成了学术诚信的基本原则，它们维护、促进的是真才实学。在这些原则的鼓励下，教师、学生都能自由、公平、公开地讨论学术观点——这才是大学教育的核心与灵魂。

第二部分
引用格式快速入门

这里将开启本书的第二部分，它涵盖了你在求学期间可能遇到的几乎所有领域的引用规范。倘若这个世界简单如一，那么所有的格式应该如出一辙。但事实上，每一种引用格式都有它自己的特点，需要顾及到细碎的注意事项。所幸每一种格式都不太复杂。

第5章是引用格式概述，介绍通用于所有领域的一些基本理念。接下来，我为每一种引用格式单独开辟一章，详细描述其风格，准确地告知你该如何引用各类著作——期刊上的文章、三卷本作品的第二卷、播客节目等都有涉及。既然各种格式是分开呈现的，实际使用时你只需参考对应的章节，即可顺利完成论文的写作。

如何选定恰当的引用格式呢？这取决于论文所属的领域、老师的要求，以及你自己的偏好。如果你的论文属于人文学科，建议使用芝加哥格式或MLA格式；如果是社会科学、工程、教育或商业领域，建议选用芝加哥格式或APA格式。生物科学、化学、计算机科学、数学和物理学都有各自的引用格式，有的时候某个领域还适用不止一种格式。我将分别予以说明，并提供大量的例证。

最后一章（第14章）是各类引用格式常见问题解答，回应了一些经常被提到的问题，比如：我读过的背景资料需要引用吗？参考文献的数量有最低要求吗？我可以在脚注里对问题展开分析吗？

现在，我们一起进入引用格式的概述部分，然后再来探讨每种格式的特征。

5 引用格式概述

注明参考资料的来源对学术诚信至关重要,实现方法是选定一种格式对它们进行合理地引用。选取何种格式由你的研究领域决定,如果你是学生,可以询问老师的建议,或结合自己的偏好,选择恰当的格式。

主流的引用格式有如下三种:

- 芝加哥(或杜拉宾式)引用格式:可用于许多领域。
- MLA 格式:适用于人文学科。
- APA 格式:适用于社会科学、教育学和商学。

其他的许多学科也形成了各自独特的引用格式:

- CSE 格式:适用于生物科学。
- AMA 格式:适用于生物医学、医学、护理学和牙医学。
- ACS 格式:适用于化学。
- 物理学的 AIP 格式,天体物理学和天文学的引用格式。
- 适用于数学、计算机科学和工程学的各类引用格式。

上述格式我都会讲到,同时附上明确的指导和充足的例证,相信能扫清你在引用规范方面的障碍。这样你就可以把更多的精力投入到论文写作中,无需为引用风格感到焦虑。不同的格式分别占据单独一章,直接翻到对应的部分学习即可。然后你就知道该如何引用图书、文章、网站、电影、音乐演出、政府公文、文字简讯、播客节目、推特帖文等内容了,几乎可以覆盖所有类型的参考资料。

为什么会生出这么多不同的引用格式呢?为什么不能在全部文章里自始至终都用同一种格式?原因是不同的学科领域不允许你如此。每个领域形成了各自的引文风格来满足自身需要,一篇德语语言研究的论文和一篇遗传学的论文就意味着两种不同的引用格式,你只能照做。例如，某些学科的格式只要求列出作者、期刊和页码信息，无需写明文章的标题。但是到了人文和社会科学领域的引用规范，文章标题就绝不能省去，这些学科的要求就是这样的。你想想看，有没有道理？

以我自己的文章为例，请比较以下三种引用格式：

芝加哥格式	Lipson, Charles. "Why Are Some International Agreements Informal?" *International Organization* 45, no. 4 (Autumn 1991): 495–538.
APA 格式	Lipson, C. (1991). Why are some international agreements informal? *International Organization*, *45*, 495–538.
ACS 格式	Lipson, C. *Int. Org.* **1991**, *45*, 495–538.

每一种都算不上复杂，只是彼此不同而已。当你走出化学实验室，加入一门研读莎士比亚的课程，就要暂时放下烧杯，以及你所熟悉的化学论文引用格式。不过，别着急，对于化学论文，只需跳至第 11 章；对于莎士比亚，翻开第 7 章，参照人文学科的 MLA 格式完成写作。章节内部以表格的形式编排，简洁直观，实例丰富，极易上手。莎士比亚在名剧《麦克白》中写道，"double, double toil and trouble（加倍又加倍，劳身又劳神）"——你手中的这本手册一定不会让你产生这种感觉。

尽管引用的风格不同，所有格式都指向相同的基本目标：

- 标示出那些引用的内容，并说明引文来源。
- 为读者提供准确的信息，以便他们自行查证。

幸运的是，不同引用格式需要的信息要素大体类似。也就是说，无论未来要用哪种体例，你都无需刻意调整笔记策略，只要从一开始把必要的信息都保存好就可以了。拿到新资料的第一时间就应该打开读书笔记，备份著录信息。如果是打印或复印的材料，尽快将出处信息写在材料的第一页上。一旦养成这样的习惯，每次都及时记录，你就不会出错。等到写论文的时候，也不会手忙脚乱。

你还极有可能用到在线的参考文献生成器。这些程序有助于收集信息和编排格式，但它们的质量良莠不齐，要尽量找更好、更可靠的工具来辅助写作。有些程序只能处理最常见的引用风格或著作类型（如果你刚好需要以 CSE 格式引用一集播客，那就只能祝你好运了）。确保文献信息准确、完整是论文写作者的责任。如有疑惑，可以查阅本书案例，对照修正细节。

参考文献的最终呈现形态与选用的引用格式有关。大部分格式都包括两部分：在论文正文中的某种标记（通常紧跟在引文之后）和文献的完整信息列表（含有更多细节，通常写在论文结尾）。文内标注可能是一个数字，以上标"99"、方括号"[99]"或圆括号"(99)"

的形式出现；也可能是放在圆括号内的一段简短说明，包含文献作者的姓氏、出版日期、页码等关键信息。更完整的参考文献列表则需要作者的姓名、著作的标题、出版日期等各种细节。

比如，芝加哥（或杜拉宾式）引用格式❶采用的是上标数字匹配注释的形式。注释可以放在每一页的底部（脚注）或文件的末尾（尾注）。脚注和尾注只有位置的差异。你可以借助文字处理软件来设定注释的位置。注释有两种风格，完整注释和简略注释。另外还有论文末尾的参考文献列表（Bibliography）。

另外两种主流格式 MLA 和 APA❷则采取不同的体例。文内夹注的形式是（Stewart 154）或（Stewart, 2018）；论文的末尾是参考文献（Reference List），列出文献的详细信息。

你所在的院系、就读的学校或合作的出版社也许偏爱某种格式，甚至会指定你使用某种格式。当然，有些时候你有权自主选择。总之，只要写论文，就需要参考文献，不妨尽早熟悉情况，在动笔前就定好体例，然后一以贯之地执行下去。

> **选择引用格式的建议：**
>
> 　　一定要先咨询每门课的任课教师，请他们为引用格式提供建议。写作时应保持格式一致。

做好引用格式，关键在于实现一致性。坚持使用相同的缩写和大写形式，不要在同一篇论文中混用各种版本。很容易发生的现象是，在一条文献中为一卷书标注了"Volume"，到另一条就缩写为"Vol."，第三次用的是"vol."。每个人都难逃此类问题，因此要仔细检查。我们还可能把"章"同时写成"chap."和"ch."。建议你在第一次做的时候就尽量统一，然后回头复查，纠正偏差。文字处理软件的查找、替换功能正好派上用场。

希望这本书可以提供一站式服务，帮助你应对所有有关引用的问题。这里涵盖了所有的体例和各式各样的被引内容。翻到每一种格式，你都能找到书籍、文章、未发表的论文、网站等信息的引用方法。对于乐谱、预印本等特殊文件，我将在谈到相应领域的时候予以介绍——物理学家经常引用预印本，但不怎么引用贝多芬，所以物理学的那一章只会反映物理学科的需求。人文学科的学生不仅引用贝多芬，还可能引用舞蹈表演、戏剧和诗歌，

---

❶ 在第 6 章你就会了解到两个名称指的是同一种格式，它们源于芝加哥大学出版社出版的两本手册。*The Chicago Manual of Style*（2017 年，第 17 版）是该格式最全面的指导手册。对于学生而言，可以参考 Kate L. Turabian 为该格式编写的简略本：Kate L. Turabian, *A Manual for Writers of Research Papers, Theses, and Dissertations*, 9th ed., rev. Wayne C. Booth, Gregory C. Colomb, Joseph M. Williams, Joseph Bizup, William T. FitzGerald, and the University of Chicago Press Editorial Staff (Chicago: University of Chicago Press, 2018).

❷ MLA 代表 Modern Language Association（美国现代语言学会），APA 代表 American Psychological Association（美国心理学会），详见第 7 章和第 8 章。

❸ Reference lists、bibliographies 等词都表示"参考文献"，但应用的场景有一些差异。在后面的章节中，我将解释其中的细节和不同格式对"参考文献"的命名方式。

具体如何操作，请阅读 MLA 格式的那一章。如果你需要引用某类冷门内容，我还会告诉你到哪儿可以获取每种引用格式的更多信息。

5.1 悬挂缩进

最后说一点共性：MLA 格式、APA 格式和其他许多学科的参考文献列表都采用一种特别的编排风格，称为"悬挂缩进"。这种排版只见于参考文献列表，不用于脚注或尾注。悬挂缩进与常规的段落缩进相反。常规缩进是将首行缩进，段落其余的行保持正常。悬挂缩进是保留第一行的正常长度，其余的全部缩进。请看示例：

Rothenberg, Gunther E. "Maurice of Nassau, Gustavus Adolphus, Raimondo Montecuccoli, and the 'Military Revolution' of the Seventeenth Century." In *Makers of Modern Strategy from Machiavelli to the Nuclear Age*, edited by Peter Paret, 32–63. Princeton, NJ: Princeton University Press, 1986.

Spooner, Frank C. *Risks at Sea: Amsterdam Insurance and Maritime Europe*, 1766–1780. Cambridge: Cambridge University Press, 1983.

采取这种不寻常的格式是有原因的。悬挂缩进能让浏览一长串参考文献的人迅速看清作者的名字。为了提醒你适应并使用这种排版，当讲到相关引用格式的时候，我将反复呈现类似的例证。（科技论文的顺序编码制是例外，这些学科的参考文献列表带有数字编号，不进行悬挂缩进。操作起来也很简单，后文我会解释。）

为了进一步突显作者信息，大多数参考文献列表会把作者的姓氏提到前面。但是，如果同一位作者重复出现，处理方式又有不同。APA 格式要求每一次都写作者的全名。芝加哥格式曾一度允许用三条英文破折号加上英文句号代替重复的作者名，但现在像 APA 格式一样，鼓励直接写出作者名字❶。MLA 格式用三条连字符加英文句号代替重复出现的作者。

Talbot, Ian. *A History of Modern South Asia: Politics, States, Diasporas.* Yale UP, 2016.

---. *Pakistan: A Modern History.* Palgrave Macmillan, 2010.

悬挂缩进也是通过文字处理软件来设置的：找到格式选项，进入段落格式，在下拉菜单中选择悬挂缩进就完成了。

---

❶ 三条英文破折号（em dash）与三条连字符（hyphen）的长度不同，在屏幕和纸质文件上的显示效果也不同，前者看起来是一条实线（———），后者则像虚线（---）。坦率讲，你不需要在论文中格外关注这个问题。如果可以的话，使用你喜欢的那个，但两者都没错。

5.2 到哪里获取更多信息

到目前为止，我已经介绍了在所有引用格式中通用的一些基本理念。当然，你心里一定还有很多疑问，或许是针对所有格式，或许针对特定的一、两种格式。这里先按下不表。我将在每一章的结尾回答当前格式涉及的常见问题，还将专门在第 14 章解答那些更宽泛的问题。

如果在某个章节末尾你没能找到疑问的答案，一定记得翻开第 14 章，应该会有新的发现。如果你的疑问超出了这本书的范围，可以找一本对应引用格式的官方指南，深入阅读。大部分引用格式都有此类书目出版了，我会在它们各自的章节罗列出来。

5.3 开始具体操作

所有信息已经分门别类地放到每种格式所在的章节中，希望可以最大限度地提高效率。

第 6 章	芝加哥（或杜拉宾式）引用格式
第 7 章	MLA 格式：人文学科
第 8 章	APA 格式：社会科学、教育学和商学
第 9 章	CSE 格式：生物科学
第 10 章	AMA 格式：生物医学、医学、护理学和牙医学
第 11 章	ACS 格式：化学
第 12 章	物理学、天体物理学和天文学的引用格式
第 13 章	数学、计算机科学和工程学的引用格式

每一章的开头，先简述当前引用格式的基本信息。接着，举例说明如何引用各类著作。图书、期刊文章等常用的参考文献在前，性质类似的文献类型归纳到一起讨论。最前面的三章是三种主流格式，篇幅最长，内容也最详实，所以我还为它们制作了索引表，方便你快速定位期刊论文、博客文章等内容的引用方式，找到相关的例子。

本书的另一特色是讲解各种在线或电子资源的引用规则。某些情况下，官方指南并不包括播客、短信这类的文献来源，要么是因为特定领域不把它们作为参考文献，要么是因为官方的风格指南近期没有更新。在可能需要引用此类信息的场景中，我根据引用格式的一般规则，尝试提供了一些例子。新类型的文献来源还会继续涌现，如果无法在书中找到实例，你可能也要像我一样，合理地预测其引用格式。

所有的引用规则都无需背诵。毕竟，这其中有太多零散的细节。需要引用论文、播客节目或任何一类著作时，打开这本书的表格查阅即可。随着实践的增多，你会越来越了解自己学科的引用格式。

　　解说完每种格式并回答相关问题之后，我在第 14 章整理了一些共性的问题，统一加以分析和回复。接下来，让我们来看看如何用每一种格式完成引用吧。

6

芝加哥（或杜拉宾式）引用格式

芝加哥大学出版社出版的 *The Chicago Manual of Style* 是芝加哥引用格式的官方指南，也是该格式的基础。截至 2017 年，这本手册已经更新至第 17 版，成为一本细致阐述参考文献、学术文体的权威著作，相关资料可到网络上获取（请访问：www.chicagomanualofstyle.org）。Kate L. Turabian（凯特·杜拉宾）编著的 *A Manual for Writers of Research Papers, Theses, and Dissertations* 是该格式的简略本，涵盖了学生论文的各个方面，目前出版到第 9 版（2018 年）。如果你想全面了解芝加哥格式的使用细节，包括一些特殊情况的处理方法，请仔细阅读本章内容。

6.1 完整注释、简略注释和参考文献

为引用内容添加注释时❶，芝加哥格式经常使用如下两种风格：

（1）首次出现时的完整注释（首见注释）+ 再出现时的简略注释

为书目、论文、文件等任何条目做首见注释时，都要提供完整的文献信息，若后续再使用同一条目，只需添加简略注释。既然首见注释已经包含了全部信息，这种风格通常不再需要参考文献列表（Bibliography）。当然，你所在的院系或合作的出版社可能仍会要求你提供参考文献列表，建议事先向他们咨询。

（2）简略注释+参考文献

所有的注释都是简略注释。完整的文献信息仅在参考文献（Bibliography）中列出。
由此可见，一个条目或许会涉及以下三种表述方法。具体书写形式请查阅本章的表格。

---

❶ 芝加哥（或杜拉宾式）引用格式同样允许使用"作者-年份"的文内夹注体系，即在圆括号内注明文献作者和出版时间，例如：(Walls 2017)，完整的文献信息出现在文后的参考文献列表内。鉴于该体例与 APA 格式相似（见第 8 章），为简化行文，本章不再赘述。

A. 首见（完整）注释
B. 简略注释
C. 参考文献

上面介绍的第一种风格需要的是 A+B 组合，第二种风格需要 B+C 组合。

这一章将告诉你如何采用芝加哥体例引用图书、在线期刊、活页乐谱、音乐录像带等各类作品。我按照字母顺序将它们整理成表格（表 6.1~表 6.4），同时附上页码信息，以便你迅速定位查看对应内容。在本章的最后，我回答了一些有关芝加哥格式的常见问题。

表6.1　芝加哥引用格式索引表

文献类型	在本书中页码	文献类型	在本书中页码
摘要（abstract）	57	电子论坛、邮件列表（electronic forum or mailing list）	73
广告（advertisement）	68–69	邮件（email）	72
档案材料（archival material）	59–60	百科全书（encyclopedia）	61
艺术作品（artwork）	65–66	脸书 Facebook	71–72
有声书（audiobook）	67	图（figure）	66
《圣经》（Bible）	62	电影（film）	65
博客（blog）		政府文献（government document）	69
评论（comment）	70	曲线图（graph）	66
文章、帖文（post）	70	照片墙（Instagram）	71–72
图书（book）		访谈（interview）	63
匿名作者（anonymous author）	54	期刊论文（journal article）	56–57
主编图书中的一章（chapter in edited book）	56	非英语（foreign language）	57
电子图书（e-book）	54	《古兰经》（Koran）	62
主编图书（edited）	53	讲座（lecture）	67
多位作者（multiple authors）	53	讲座的录音（recording）	67
多个版本（multiple editions）	53	杂志的文章（magazine article）	58
多卷本著作（multivolume work）	55	地图（map）	66
无作者（no author）	54	多媒体应用（multimedia app）	71
单一作者（one author）	52	音乐（music）	
在线图书（online）	54	唱片内页文字说明（liner notes）	68
早期版本的重印版（reprint）	55–56	音乐唱片（recording）	66–67
多本图书，同一作者（several by same author）	52	活页乐谱（sheet music）	68
多卷本中的单卷作品（single volume in a multivolume work）	55	音乐录像带（video）	68
译著（translated volume）	56	报纸的文章（newspaper article）	58
图表（chart）	66	未发表的论文（paper, unpublished）	58–59
古典著作（classical work）	62	私人通信（personal communication）	63
舞蹈演出（dance performance）	64		
词典（dictionary）	61		
私信（direct message）	72	照片（photograph）	66
学位论文（dissertation）	58–59		

续表

文献类型	在本书中页码	文献类型	在本书中页码
戏剧（play） 　戏剧演出（performance） 　戏剧剧本（text）	 64 64	表格（table）	66
		电视节目（television program）	64–65
		短信（text message）	72
播客（podcast）	71	毕业论文（thesis）	58–59
诗歌（poem）	63–64	推特（Twitter）	71–72
预印本或未定稿文件（preprint or working paper）	59	视频（video）	70–71
		电子游戏（video game）	71
书评（review）	58	网站、网页（website or page）	69–70
社交媒体（social media）	72		
演讲、报告或讲座（speech, lecture, or talk） 朗读、讲座或有声书的录音（recording）	62–63 67		

表6.2　芝加哥引用格式：注释和参考文献

图书，单一作者	首见注释	⁹⁹ Charles Lipson, *Reliable Partners: How Democracies Have Made a Separate Peace* (Princeton, NJ: Princeton University Press, 2003), 22–23. ▶ 这是第 99 条注释的内容，引文出自原著第 22–23 页 ▶ 脚注和尾注不使用悬挂缩进。只有参考文献（Bibliography）才用 ⁹⁹ Terry McDermott, *Off Speed: Baseball, Pitching, and the Art of Deception* (New York: Pantheon, 2017), 1028. ▶ 芝加哥格式允许缩写连续的页码，规则是：如果起始页的页码不超过两位数，应列出所有数字（1–12, 44–68）；如果起始页是三位数或以上，并且起始数字不是 100 的倍数，则省略重复的部分（203–6, 345–52, 349–402，例外的是 200–223）。看起来有点复杂，实际上也确实不容易理解。好消息是，把所有数字都写出来，不作任何省略，也没问题
	简略注释	⁹⁹ Lipson, *Reliable Partners*, 22–23. ⁹⁹ McDermott, *Off Speed*, 102–8. ▶ 如有可能，将标题缩写至四个词以内
	参考文献	Lipson, Charles. *Reliable Partners: How Democracies Have Made a Separate Peace.* Princeton, NJ: Princeton University Press, 2003. McDermott, Terry. *Off Speed: Baseball, Pitching, and the Art of Deception.* New York: Pantheon, 2017.
多本图书，同一作者	首见注释	⁹⁹ Ian Talbot, *A History of Modern South Asia: Politics, States, Diasporas* (New Haven, CT: Yale University Press, 2016). ¹⁰⁰ Ian Talbot, *Pakistan: A Modern History* (London: Palgrave Macmillan, 2010).
	简略注释	⁹⁹ Talbot, *History of Modern South Asia.* ¹⁰⁰ Talbot, *Pakistan.*
	参考文献	Talbot, Ian. *A History of Modern South Asia: Politics, States, Diasporas.* New Haven, CT: Yale University Press, 2016. Talbot, Ian. *Pakistan: A Modern History.* London: Palgrave Macmillan, 2010. ▶ 建议在参考文献中把重复的作者姓名照常写出来。虽然它们可以用三条英文破折号加英文句号来代替，但是相应条目中就看不到姓名这一重要信息了，所以不推荐这种做法。如果不知道如何输入三条英文破折号，可直接用三个连字符 ▶ 同一作者名下的著作按标题的首字母顺序排列。标题开头的冠词（A/An/The）不作为排序依据

续表

图书，多位作者	首见注释	[99] John Martin Fischer and Benjamin Mitchell-Yellin, *Near-Death Experiences: Understanding Visions of the Afterlife* (Oxford: Oxford University Press, 2016), 15–18. ▶ 对于四位或更多的作者来说，仅标注第一位作者，其后注上"et al."（源自拉丁语 *et alii*，表示"以及其他人"）。假设上例有超过四位作者，那么要写作：John Martin Fischer et al., *Near-Death …*
	简略注释	[99] Fischer and Mitchell-Yellin, *Near-Death Experiences*, 15–18. ▶ 四个单词以内的标题无需缩写
	参考文献	Fischer, John Martin, and Benjamin Mitchell-Yellin. *Near-Death Experiences: Understanding Visions of the Afterlife*. Oxford: Oxford University Press, 2016. ▶ 只有第一个作者的姓氏提到了名字前面 ▶ 这里最多列出十位共同作者。如果合著者超过十名，请列出前七个，然后加上"et al."字样
图书，多个版本	首见注释	[99] Annelise Orleck, *Common Sense and a Little Fire: Women and Working-Class Politics in the United States, 1900–1965*, 2nd ed. (Chapel Hill: University of North Carolina Press, 2017). [99] William Strunk Jr. and E. B. White, *The Elements of Style*, 50th anniversary ed. (New York: Longman, 2009), 12.
	简略注释	[99] Orleck, *Common Sense*. [99] Strunk and White, *Elements of Style*, 12. ▶ 为使注释更加简洁，省去了标题开头的冠词（完整标题为 *The Elements of Style*）和著作的版次
	参考文献	Orleck, Annelise. *Common Sense and a Little Fire: Women and Working-Class Politics in the United States, 1900–1965*, 2nd ed. Chapel Hill: University of North Carolina Press, 2017. Strunk, William, Jr., and E. B. White. *The Elements of Style*. 50th anniversary ed. New York: Longman, 2009.
主编图书	首见注释	[99] Annie Bunting and Joel Quirk, eds., *Contemporary Slavery: Popular Rhetoric and Political Practice* (Vancouver: University of British Columbia Press, 2017). [99] Jeffrey Ian Ross, ed., *Routledge Handbook of Graffiti and Street Art* (New York: Routledge, 2016). [99] Mirjam Künkler, John Madeley, and Shylashri Shankar, eds., *A Secular Age beyond the West* (Cambridge: Cambridge University Press, 2017).
	简略注释	[99] Bunting and Quirk, *Contemporary Slavery*. ▶ 此处不需要用 editor 的缩写 ed.标注编者 [99] Ross, *Graffiti and Street Art*. ▶ 选取最有代表性的词来缩写标题 [99] Künkler, Madeley, and Shankar, *Secular Age*.
	参考文献	Bunting, Annie, and Joel Quirk, eds. *Contemporary Slavery: Popular Rhetoric and Political Practice*. Vancouver: University of British Columbia Press, 2017. Ross, Jeffrey Ian, ed. *Routledge Handbook of Graffiti and Street Art*. New York: Routledge, 2016. Künkler, Mirjam, John Madeley, and Shylashri Shankar, eds. *A Secular Age beyond the West*. Cambridge: Cambridge University Press, 2017.

续表

图书，匿名作者或无作者	首见注释	⁹⁹ Anonymous, *The Secret Lives of Teachers* (Chicago: University of Chicago Press, 2015). ⁹⁹ *Golden Verses of the Pythagoreans* (Whitefish, MT: Kessinger, 2003).
	简略注释	⁹⁹ Anonymous, *Secret Lives of Teachers*. ⁹⁹ *Golden Verses of Pythagoreans*.
	参考文献	Anonymous. *The Secret Lives of Teachers*. Chicago: University of Chicago Press, 2015. *Golden Verses of the Pythagoreans*. Whitefish, MT: Kessinger, 2003. ▶ 如果原著作者明确写作 Anonymous，则应在参考文献内如实记录。如果原著未列出作者姓名，你可以直接以标题开头，也可以添加"anonymous"一词
电子图书	首见注释	⁹⁹ Gary Taubes, *The Case against Sugar* (New York: Alfred A. Knopf, 2016), Kindle. ⁹⁹ Kate Auspitz, *Wallis's War: A Novel of Diplomacy and Intrigue* (Chicago: University of Chicago Press, 2015), iBooks. ▶ 应注明阅读电子书使用的设备或应用。如果需要指出文件是 EPUB 还是 PDF 等其他格式，可以写在设备或应用信息的后面（例如，Adobe Digital Editions EPUB） ▶ 指明引文具体位置时，最好借助章节编号、小标题等标志。电子书页码会因为使用了不同设备或应用程序而发生变化
	简略注释	⁹⁹ Taubes, *Case against Sugar*, chap. 2. ⁹⁹ Auspitz, *Wallis's War*.
	参考文献	Taubes, Gary. *The Case against Sugar*. New York: Alfred A. Knopf, 2016. Kindle. Auspitz, Kate. *Wallis's War: A Novel of Diplomacy and Intrigue*. Chicago: University of Chicago Press, 2015. iBooks. Adobe Digital Editions EPUB.
在线图书	首见注释	⁹⁹ Charles Dickens, *Great Expectations* (1867; Project Gutenberg, 2008), chap. 2, http://www.gutenberg.org/files/1400/1400-h/1400-h.htm. ⁹⁹ Frederick Jackson Turner, *The Frontier in American History* (New York: Henry Holt, 1921), 6, https://books.google.com/books/about/The_Frontier_in_American_History.html?id=vtF1AAAAMAAJ. ▶ 应将网址写在最后。推荐使用基于 DOI 的访问地址（DOI 编码前面添加 https://doi.org/） ▶ 在线图书也有不同的格式。如果可能，请引用扫描版，而不是 HTML 或 EPUB 等自适应式的格式。扫描版的格式稳定，更加适合引用页码。上例中 Turner 的著作就是原版的扫描版 ▶ 如果手中可用的版本都无法对应到准确的页码，可以使用章节编号、小标题等类似的标志
	简略注释	⁹⁹ Dickens, *Great Expectations*, chap. 2. ⁹⁹ Turner, *Frontier in American History*, 6.
	参考文献	Dickens, Charles. *Great Expectations*. 1867. Project Gutenberg, 2008. http://www.gutenberg.org/files/1400/1400-h/1400-h.htm. Turner, Frederick Jackson. *The Frontier in American History*. New York: Henry Holt, 1921. https://books.google.com/books/about/The_Frontier_in_American_History.html?id=vtF1AAAAMAAJ.

多卷本著作	首见注释	⁹⁹ Otto Pflanze, *Bismarck and the Development of Germany*, 3 vols. (Princeton, NJ: Princeton University Press, 1963–90), 1:153. ⁹⁹ Babette Rothschild, *The Body Remembers*, 2 vols. (New York: W. W. Norton, 2017), 2:16.
	简略注释	⁹⁹ Pflanze, *Bismarck*, 1:153. ⁹⁹ Rothschild, *Body Remembers*, 2:16.
	参考文献	Pflanze, Otto. *Bismarck and the Development of Germany*. 3 vols. Princeton, NJ: Princeton University Press, 1963–90. Rothschild, Babette. *The Body Remembers*, 2 vols. New York: W. W. Norton, 2017.
多卷本中的单卷作品	首见注释	⁹⁹ Robert A. Caro, *The Years of Lyndon Johnson*, vol. 4, *The Passage of Power* (New York: Alfred A. Knopf, 2012), 237. ⁹⁹ Bruce E. Johansen, *Global Warming in the 21st Century*, vol. 2, *Melting Ice and Warming Seas* (Westport, CT: Praeger, 2006), 71. ⁹⁹ Akira Iriye, *The Globalizing of America*, vol. 3 of *Cambridge History of American Foreign Relations*, ed. Warren I. Cohen (Cambridge: Cambridge University Press, 1993), 124. ▶ 可以看出，Caro 是整套书的作者。Cohen 是整套书的主编，Iriye 是第三卷书的作者
	简略注释	⁹⁹ Caro, *Years of Lyndon Johnson*, 4:237. ▶ 或 ⁹⁹ Caro, *Passage of Power*, 237. ⁹⁹ Johansen, *Global Warming*, 2:71. ▶ 或 ⁹⁹ Johansen, *Melting Ice and Warming Seas*, 71. ⁹⁹ Iriye, *Globalizing of America*, 124.
	参考文献	Caro, Robert A. *The Years of Lyndon Johnson*. Vol. 4, *The Passage of Power*. New York: Alfred A. Knopf, 2012. Johansen, Bruce E. *Global Warming in the 21st Century*. Vol. 2, *Melting Ice and Warming Seas. Westport, CT: Praeger, 2006.* Iriye, Akira. *The Globalizing of America*. Vol. 3 of *Cambridge History of American Foreign Relations*, edited by Warren I. Cohen. Cambridge: Cambridge University Press, 1993.
早期版本的重印版	首见注释	⁹⁹ Jacques Barzun, *Simple and Direct: A Rhetoric for Writers*, rev. ed. (1985; repr., Chicago: University of Chicago Press, 1994), 27. ⁹⁹ Adam Smith, *An Inquiry into the Nature and Causes of the Wealth of Nations* (1776), ed. Edwin Cannan (Chicago: University of Chicago Press, 1976). ▶ 1776 年紧跟在书名后出现，标示 Smith 原著的出版时间。新的版本经 Edwin Cannan 编辑而成。相比之下，Barzun 的作品是单纯的重印，因此原始出版时间也写在了出版信息的部分
	简略注释	⁹⁹ Barzun, *Simple and Direct*, 27. ⁹⁹ Smith, *Wealth of Nations*, vol. Ⅰ, bk. Ⅳ, chap. Ⅱ: 477. ▶ 重印出版的 Smith 作品只有一卷，但上面的示例仍保留了 1776 原版作品的卷数信息。只写页码，不写卷数、章节，也是允许的，但是把出处写完整能帮助拿到了不同版本的读者

续表

早期版本的重印版	参考文献	Barzun, Jacques. *Simple and Direct: A Rhetoric for Writers*. 1985. Reprint, Chicago: University of Chicago Press, 1994. Smith, Adam. *An Inquiry into the Nature and Causes of the Wealth of Nations*. 1776. Edited by Edwin Cannan. Chicago: University of Chicago Press, 1976.
译著	首见注释	[99] Hesiod, *The Poems of Hesiod: Theogony, Works and Days, & The Shield of Herakles*, trans. Barry B. Powell (Oakland: University of California Press, 2017), 30–35. [99] Alexis de Tocqueville, *Democracy in America* (1835), ed. J. P. Mayer, trans. George Lawrence (New York: HarperCollins, 2000). ▶ 译者和编者按其在图书扉页上出现的顺序排列 [99] Seamus Heaney, trans., *Beowulf: A New Verse Translation* (New York: Farrar, Straus and Giroux, 2000). ▶ 对于 *Beowulf*（《贝奥武夫》），译者的名字写在了书名之前，原因是扉页上只有 Seamus Heaney 这个名字（该诗的作者是无名氏）。如果扉页上只有编辑或编纂者的名字，处理方法同上
译著	简略注释	[99] Hessiod, *Poems*, 30–35. [99] Tocqueville, *Democracy in America*. [99] *Beowulf*. ▶ 或 [99] Heaney, *Beowulf*.
译著	参考文献	Hesiod. *The Poems of Hesiod: Theogony, Works and Days, & The Shield of Herakles*. Translated by Barry B. Powell. Oakland: University of California Press, 2017. Tocqueville, Alexis de. *Democracy in America*. 1835. Edited by J. P. Mayer. Translated by George Lawrence. New York: HarperCollins, 2000. Heaney, Seamus, trans. *Beowulf: A New Verse Translation*. New York: Farrar, Straus and Giroux, 2000.
主编图书中的一章	首见注释	[99] Benjamin J. Cohen, "The Macrofoundations of Monetary Power," in *International Monetary Power*, ed. David M. Andrews (Ithaca, NY: Cornell University Press, 2006), 31–50.
主编图书中的一章	简略注释	[99] Cohen, "The Macrofoundations of Monetary Power," 31–50.
主编图书中的一章	参考文献	Cohen, Benjamin J. "The Macrofoundations of Monetary Power." In *International Monetary Power*, edited by David M. Andrews, 31–50. Ithaca, NY: Cornell University Press, 2006.
期刊论文	首见注释	[99] Stefan Timmermans, "Matching Genotype and Phenotype: A Pragmatist Semiotic Analysis of Clinical Exome Sequencing," *American Journal of Sociology* 123, no. 1 (July 2017): 137. [99] Smita Sahgal, "Situating Kingship within an Embryonic Frame of Masculinity in Early India," *Social Scientist* 43, no. 11/12 (November-December 2015): 5, http://www.jstor.org.pnw.idm.oclc.org/stable/24642382. [99] Krista A. Capps, Carla L. Atkinson, and Amanda T. Rugenski, "Implications of Species Addition and Decline for Nutrient Dynamics in Fresh Waters," *Freshwater Science* 34, no. 2 (June 2015): 490, https://doi.org/10.1086/681095. ▶ 对于在线访问的期刊论文，要在最后注明网址或 DOI 编码。基于 DOI 的访问地址（DOI 编码前面添加 https://doi.org/）比普通的网页地址更合适。如果无法提供 DOI 或网址，要列出数据库的名称

续表

	首见注释	▶ 若作者人数为四位或以上，仅标注第一位作者，其后写上"et al." ▶ 可以用同样的方式引用某篇论文的摘要，但是要在论文的标题后面加上"abstract"一词
期刊论文	简略注释	99 Timmermans, "Matching Genotype and Phenotype," 137. 99 Sahgal, "Situating Kingship," 5. 99 Capps, Atkinson, and Rugenski, "Implications of Species Addition and Decline," 490. ▶ 注释中标出的是引文的页码范围（如果能提供具体页码的话）。参考文献中标出的是整篇论文的页码范围
	参考文献	Timmermans, Stefan. "Matching Genotype and Phenotype: A Pragmatist Semiotic Analysis of Clinical Exome Sequencing." *American Journal of Sociology* 123, no. 1 (July 2017): 137–77. Sahgal, Smita. "Situating Kingship within an Embryonic Frame of Masculinity in Early India." *Social Scientist* 43, no. 11/12 (November-December 2015): 3–26. http://www.jstor. org.pnw.idm.oclc.org/stable/24642382. Capps, Krista A., Carla L. Atkinson, and Amanda T. Rugenski. "Implications of Species Addition and Decline for Nutrient Dynamics in Fresh Waters." *Freshwater Science* 34, no. 2 (June 2015): 485–96, https://doi.org/10.1086/681095. ▶ 只有第一个作者的姓氏提到了名字前面 ▶ 这里最多列出十位共同作者。如果合著者超过十名，请列出前七个，然后加上"et al."
期刊论文，非英语	首见注释	99 Zvi Uri Ma'oz, "Y a-t-il des juifs sans synagogue?," *Revue des Études Juives* 163 (juillet-décembre 2004): 485. ▶ 或 99 Zvi Uri Ma'oz, "Y a-t-il des juifs sans synagogue?" [Are there Jews without a synagogue?], *Revue des Études Juives* 163 (juillet-décembre 2004): 485.
	简略注释	99 Ma'oz, "Y a-t-il des juifs sans synagogue?," 485.
	参考文献	Ma'oz, Zvi Uri. "Y a-t-il des juifs sans synagogue?" *Revue des Études Juives* 163 (juillet-décembre 2004): 483–93. ▶ 或 Ma'oz, Zvi Uri. "Y a-t-il des juifs sans synagogue?" [Are there Jews without a synagogue?]. *Revue des Études Juives* 163 (juillet-décembre 2004): 483–93.
报纸或杂志的文章	首见注释	99 "Retired U.S. General Is Focus of Inquiry over Iran Leak," *New York Times*, June 28, 2013, A18. ▶ 此处指的是报纸的第 A18 版。由于许多报纸有不同的版本，且分版面的方式不同，你也可以选择省略版面信息 99 Shivani Vora, "Why Your Airline Says It's Sorry," *New York Times*, July 30, 2017, Sunday edition. 99 Evan Halper, "Congress Takes Aim at the Clean Air Act, Putting the Limits of California's Power to the Test," *Los Angeles Times*, August 3, 2017, 3:00 a.m. PDT, http://www.latimes.com/politics/la-na-pol-smog-republicans-20170803-story.html. ▶ 对于在线访问的文章，要在最后注明网址或 DOI 编码 ▶ 如果文章经常更新，请附上文章的发布时间（时间戳） 99 Ian Parker, "The Greek Warrior: How a Radical Finance Minister Took on Europe—and Failed," *New Yorker*, August 3, 2015, 46.

续表

报纸或杂志的文章	首见注释	▶ 面向普通读者的杂志、报纸上的文章，都是只按日期引用。有些杂志更像专业刊物。如果你不确定自己引用的是杂志文章还是期刊论文——看能否找到卷号。若发现了明显的卷号，就按期刊的形式来引用
	简略注释	[99] "Focus of Inquiry," *New York Times*, A18. [99] Vora, "Airline Says It's Sorry." [99] Harper, "Congress Takes Aim." [99] Parker, "The Greek Warrior," 46. ▶ 鉴于报纸和杂志通常不列入参考文献，建议你采用完整注释的形式
	参考文献	▶ 报纸、杂志的文章不列入参考文献，除非你认为某篇文章极其重要 "Retired U.S. General Is Focus of Inquiry over Iran Leak." *New York Times*. June 28, 2013, A18. Parker, Ian. "The Greek Warrior: How a Radical Finance Minister Took on Europe—and Failed." *New Yorker*, August 3, 2015, 44–57.
书评	首见注释	[99] Allison J. Pugh, review of *The Gender Trap: Parents and the Pitfalls of Raising Boys and Girls*, by Emily W. Kane, *American Journal of Sociology* 119, no. 6 (May 2014): 1773–75. [99] Giovanni Vimercati, "Soviet Pseudoscience: The History of Mind Control," review of *Homo Sovieticus: Brain Waves, Mind Control, and Telepathic Destiny*, by Wladimir Velminski, *Los Angeles Review of Books*, August 20, 2017, https://lareviewofbooks.org/article/soviet- pseudoscience-the-history-of-mind-control.
	简略注释	[99] Pugh, review of *Gender Trap*. [99] Vimercati, "Soviet Pseudoscience." ▶ 或 [99] Vimercati, review of *Homo Sovieticus*.
	参考文献	Pugh, Allison J. Review of *The Gender Trap: Parents and the Pitfalls of Raising Boys and Girls*, by Emily W. Kane. *American Journal of Sociology* 119, no. 6 (May 2014): 1773–75. Ferguson, Niall. "Ameliorate, Contain, Coerce, Destroy." Review of *The Utility of Force: The Art of War in the Modern World*, by Rupert Smith. *New York Times Book Review*, February 4, 2007, 14–15.
未发表的论文、毕业论文或学位论文	首见注释	[99] Shaojie Tang, "Profit Driven Team Grouping in Social Networks" (paper presented at the Thirty-First AAAI Conference on Artificial Intelligence, February 4-9, 2017), https://aaai.org/ocs/index.php/AAAI/AAAI17/ paper/view/ 14791/13741. [99] Lance Noble, "One Goal, Multiple Strategies: Engagement in Sino-American WTO Accession Negotiations" (master's thesis, University of British Columbia, 2006), 15. [99] Luis Manuel Sierra, "Indigenous Neighborhood Residents in the Urbanization of La Paz, Bolivia, 1910–1950," (PhD diss., State University of New York at Binghamton, 2013), ProQuest. ▶ 如果引用的论文出自商业数据库，应注明数据库的名称。如果论文需要在线查看，要标注网址
	简略注释	[99] Tang, "Profit Driven Team Grouping." [99] Noble, "One Goal, Multiple Strategies." [99] Sierra, "Indigenous Neighborhood Residents."

续表

未发表的论文、毕业论文或学位论文	参考文献	Tang, Shaojie. "Profit Driven Team Grouping in Social Networks." Paper presented at the Thirty-First AAAI Conference on Artificial Intelligence, February 4–9, 2017, https://aaai.org/ocs/index.php/AAAI/AAAI17/paper/view/14791/13741. Noble, Lance. "One Goal, Multiple Strategies: Engagement in Sino- American WTO Accession Negotiations." Master's thesis, University of British Columbia, 2006. Sierra, Luis Manuel. "Indigenous Neighborhood Residents in the Urbanization of La Paz, Bolivia, 1910–1950." PhD diss., State University of New York at Binghamton, 2013. ProQuest (3612828).
预印本或未定稿文件	首见注释	[99] John C. Rodda et al., "A Comparative Study of the Magnitude, Frequency and Distribution of Intense Rainfall in the United Kingdom," preprint, October 9, 2009, http://precedings.nature.com/documents/3847/version/1. [99] Daniel P. Gross, "The Ties That Bind: Railroad Gauge Standards and Internal Trade in the 19th Century U.S." (working paper, Berkeley Economic History Lab, University of California, Berkeley, March 2016), http://behl.berkeley.edu/working-papers/.
	简略注释	[99] Rodda et al., "Rainfall in the United Kingdom." [99] Gross, "The Ties That Bind."
	参考文献	Rodda, John C., Max A. Little, Harvey J. E. Rodda, and Patrick E. McSharry. "A Comparative Study of the Magnitude, Frequency and Distribution of Intense Rainfall in the United Kingdom." Preprint, October 9, 2009. http://precedings.nature.com/ documents/3847/version/1. Gross, Daniel P. "The Ties That Bind: Railroad Gauge Standards and Internal Trade in the 19th Century U.S." Working paper, Berkeley Economic History Lab, University of California, Berkeley, March 2016. http://behl.berkeley.edu/working-papers/.
档案材料、手稿集（纸质版、在线版）	首见注释	[99] Isaac Franklin to R. C. Ballard, February 28, 1831, series 1.1, folder 1, Rice Ballard Papers, Southern Historical Collection, Wilson Library, University of North Carolina, Chapel Hill. ▶ 各项信息的顺序安排如下： ① 作者和内容简述 ② 日期（如果有的话） ③ 文件或手稿的编号 ④ 作品系列或案卷的名称 ⑤ 图书馆（或存放地点）名称和位置；对于大众熟知的图书馆和档案馆，位置信息可省略 [99] Mary Swift Lamson, "An Account of the Beginning of the B.Y.W.C.A.," MS, [n.d.], and accompanying letter, 1891, series I, I-A-2, Boston YWCA Papers, Schlesinger Library, Radcliffe Institute for Advanced Study, Harvard University. ▶ MS=manuscript=papers（MS 的复数表示为 MSS） [99] Sigismundo Taraval, journal recounting Indian uprisings in Baja California [handwritten MS], 1734-37, ¶ 23, Edward E. Ayer Manuscript Collection no. 1240, Newberry Library, Chicago. ▶ 这份记录含有段落编号。页码、段落或其他标识符能有效帮助读者

续表

档案材料、手稿集（纸质版、在线版）	首见注释	⁹⁹ Horatio Nelson Taft, diary, February 20, 1862, p. 149 (vol. 1, January 1, 1861-April 11, 1862), Manuscript Division, Library of Congress, http://memory.loc.gov/ammem/tafthtml/tafthome.html. ⁹⁹ Henrietta Szold to Rose Jacobs, February 3, 1932, reel 1, book 1, Rose Jacobs-Alice L. Seligsberg Collection, Judaica Microforms, Brandeis Library, Waltham, MA. ▶ 缩写：如需反复提及文集或档案馆的名称，可以在第一次使用后对其进行缩写 ⁹⁹ Henrietta Szold to Rose Jacobs, March 9, 1936, A/125/112, Central Zionist Archives, Jerusalem (hereafter cited as CZA). ¹⁰⁰ Szold to Eva Stern, July 27, 1936, A/125/912, CZA.
	简略注释	⁹⁹ Isaac Franklin to R. C. Ballard, February 28, 1831, series 1.1, folder 1, Rice Ballard Papers. ▶ 简略注释的最终样式跟被引档案材料的性质有关。主要原则是方便读者，还要看一下附近的注释中是否已经有了完整的信息 ⁹⁹ Mary Swift Lamson, "Beginning of the B.Y.W.C.A.," MS [1891], Boston YWCA Papers, Schlesinger Library. ⁹⁹ Sigismundo Taraval, journal recounting Indian uprisings in Baja California, Edward E. Ayer Manuscript Collection, Newberry Library. ▶ 或 ⁹⁹ Taraval, journal, Ayer MS Collection, Newberry Library. ⁹⁹ Horatio Nelson Taft, diary, February 20, 1862, 149. ⁹⁹ Henrietta Szold to Rose Jacobs, February 3, 1932, reel 1, book 1, Rose Jacobs-Alice L. Seligsberg Collection. ¹⁰⁰ Szold to Jacobs, March 9, 1936, A/125/112, CZA. ¹⁰¹ Szold to Eva Stern, July 27, 1936, A/125/912, CZA.
	参考文献	Rice Ballard Papers. Southern Historical Collection. Wilson Library. University of North Carolina, Chapel Hill. ▶ 在脚注和尾注中，具体的档案项目通常列在首位，它是注释中最重要的内容（比如：Isaac Franklin to R. C. Ballard, February 28, 1831）。但在参考文献部分，通常先列出档案的案卷，因为此时这条信息更为重要。参考文献列表可以不体现某个单独的档案项目，除非它是整部案卷中用到的唯一素材 Boston YWCA Papers. Schlesinger Library. Radcliffe Institute for Advanced Study, Harvard University. ▶ 或 Lamson, Mary Swift. "An Account of the Beginning of the B.Y.W.C.A." MS [n.d.] and accompanying letter, 1891. Boston YWCA Papers. Schlesinger Library. Radcliffe Institute for Advanced Study, Harvard University. ▶ 如果 Lamson 的叙述是整套手稿中唯一引用到论文中的内容，则应将它的具体信息写入参考文献 Ayer, Edward E. Manuscript collection. Newberry Library, Chicago, IL. Taft, Horatio Nelson. Diary. Vol. 1, January 1, 1861-April 11, 1862. Manuscript Division, Library of Congress. http://memory.loc.gov/ammem/tafthtml/tafthome.html. Rose Jacobs-Alice L. Seligsberg Collection. Judaica Microforms. Brandeis Library, Waltham, MA. Central Zionist Archives, Jerusalem.

		续表
百科全书、词典	首见注释	⁹⁹ *Encyclopaedia Britannica*, 15th ed. (1974), s.vv. "Balkans: History," "World War Ⅰ." ⁹⁹ *Merriam-Webster*, s.v. "app (n.)," accessed April 6, 2016, http://www.merriam-webster. com/dictionary/app. ▶ 缩写 s.v.（*sub verbo*）表示 under the word（在……词条下，复数形式为 s.vv.） ▶ 如果从印刷出版的材料中引用，必须注明版本。对于 *Encyclopaedia Britannica*（《大英百科全书》）这样众所周知的参考材料，可以省略出版商、出版地和页码信息 ⁹⁹ *Encyclopaedia Britannica Online*, s.v. "Balkans," accessed September 7, 2016, http://www.britannica.com/EBchecked/topic/50325/Balkans. ⁹⁹ Dictionary.com, s.v. "metropolitan," accessed August 4, 2017, http://www.dictionary.com/ browse/metropolitan?s=t. ▶ 一些百科全书会为每篇文章附上一个访问地址，建议用这个链接表示你的文献来源，不要使用浏览器地址栏显示的网址 ⁹⁹ George Graham, "Behaviorism," in *Stanford Encyclopedia of Philosophy*, article published May 26, 2000, revised July 30, 2007, http://plato.stanford. edu/entries/behaviorism/. ▶ 或 ⁹⁹ *Stanford Encyclopedia of Philosophy*, "Behaviorism," by George Graham, article published May 26, 2000, revised July 30, 2007, http://plato.stanford.edu/entries/behaviorism/. ⁹⁹ Wikipedia, s.v. "Sufjan Stevens," last modified July 31, 2017, 15:30, https://en.wikipedia.org/wiki/Sufjan_Stevens. ▶ 如果文章含有发表日期，则应如实标注，也可以引用文章的最后修改日期。如果两者都没有，要注明访问日期 ▶ 如果引用的是有作者的条目，则可以注明作者，格式仿照多位作者所著的图书来写
	简略注释	⁹⁹ *Encyclopaedia Britannica*, s.v."World War Ⅰ." ⁹⁹ *Merriam-Webster's*, s.v."chronology." ⁹⁹ *Encyclopaedia Britannica Online,* s.v. "Balkans." ⁹⁹ Dictionary.com, s.v. "metropolitan." ⁹⁹ Graham, "Behaviorism." ▶ 或 ⁹⁹ *Stanford Encyclopedia*, "Behaviorism." ⁹⁹ Wikipedia, "Sufjan Stevens."
	参考文献	▶ 著名的百科全书和词典通常不列入参考文献。但是如果你希望引用某位作者贡献的专业文章，可以将它写在参考文献列表中 Graham, George. "Behaviorism." In *Stanford Encyclopedia of Philosophy*. Article published May 26, 2000. Revised July 30, 2007. http://plato.stanford.edu/ entries/behaviorism/. ▶ 或` *Stanford Encyclopedia of Philosophy*. "Behaviorism," by George Graham. Article published May 26, 2000. Revised July 30, 2007. http://plato.stanford.edu/entries/behaviorism/.

《圣经》《古兰经》	首见注释	⁹⁹ Genesis 1:1, 1:3–5, 2:4. ⁹⁹ Genesis 1:1, 1:3–5, 2:4 (New Revised Standard Version). ▶ 圣经的每卷书可以缩写（例如，创世记 Gen. 1:1） ▶ 接下来的四卷书缩写为 Exod.（出埃及记）、Lev.（利未记）、Num.（民数记）以及 Deut.（申命记）。如果你需要查询更多信息，在网络中搜索 "abbreviations + Bible" 即可
	简略注释	⁹⁹ Koran 18:65–82. ⁹⁹ Genesis 1:1, 1:3–5, 2:4. ⁹⁹ Koran 18:65–82.
	参考文献	▶ 对《圣经》《古兰经》等宗教典籍的引用通常不包括在参考文献中，除非你想表明你使用了某个特定版本或译本。比如： *Tanakh: The Holy Scriptures: The New JPS Translation according to the Traditional Hebrew Text*. Philadelphia: Jewish Publication Society, 1985. ▶ 省略宗教典籍的原作者姓名
古典著作	首见注释	▶ 一般情况下，如需引用希腊语、拉丁语的古典著作，作者会在正文或注释中加以说明，因此也就无需将它们列入参考文献，除非是为了突显特定的译本或某位现代作者的评述。请看一个文内引用的案例： In Pericles' Funeral Oration (2.34–46), Thucydides gives us one of history's most moving speeches. ▶ 如果写成注释的形式，可以参考如下写法： ⁹⁹ Plato, *The Republic*, trans. R. E. Allen (New Haven, CT: Yale University Press, 2006). ⁹⁹ Virgil, *The Aeneid*, trans. David Ferry (Chicago: University of Chicago Press, 2017).
	简略注释	⁹⁹ Plato, *Republic* 3.212b–414b. ⁹⁹ Virgil, *Aeneid* 5.6–31.
	参考文献	▶ 古典希腊和拉丁语作品通常不包括在参考文献列表中，除非参考的是特定译本或现代作者的评注本 Plato. *The Republic*. Translated by R. E. Allen. New Haven, CT: Yale University Press, 2006. Virgil. *The Aeneid*. Translated by David Ferry. Chicago: University of Chicago Press, 2017.
演讲、学术报告或课程讲座	首见注释	⁹⁹ Alison Dagnes, "Circles of Influence: How the Current Partisan Media System Divides Us" (paper presented at the annual meeting of the Northeastern Political Science Association, November 12, 2015). ⁹⁹ Gary Sick, lecture on US policy toward Iran (U.S. Foreign Policy Making in the Persian Gulf, course taught at Columbia University, New York, March 22, 2007). ▶ Sick 教授讲座的标题没有大写或加引号，因为它是一次普通的课程讲座，没有具体的标题。上例为这次讲座增加了一些修饰语，但你可以将其省略，直接写 lecture，即：Gary Sick, untitled lecture (U.S. Foreign …)
	简略注释	⁹⁹ Dagnes, "Circles of Influence." ▶ 或者，如果论文引用了 Skorton 不同时期发表的同主题演讲，可以在注释中标注时间，以示区分： ⁹⁹ Skorton, "State of the University Speech," 2006. ⁹⁹ Sick, lecture on US policy toward Iran.

演讲、学术报告或课程讲座	参考文献	Dagnes, Alison. "Circles of Influence: How the Current Partisan Media System Divides Us." Paper presented at the annual meeting of the Northeastern Political Science Association, November 12, 2015. Sick, Gary. Lecture on US policy toward Iran. U.S. Foreign Policy Making in the Persian Gulf, course taught at Columbia University, New York, March 22, 2007.
私人通信或访谈	首见注释	[99] Robert Alter, personal interview, October 21, 2015. [99] Nicolas Sarkozy, telephone interview, May 5, 2013. [99] George Lucas, video interview (Skype), October 27, 2016. [99] Anonymous US soldier recently returned from Afghanistan, interview by author, February 19, 2017. [99] Discussion with senior official at Department of Homeland Security, Washington, DC, January 7, 2010. ▶ 在某些情况下，你可能不希望透露采访或谈话的信息来源，或者你已提前答应受访者为其身份保密。此时，你需要做的是：a.尽量提供更多的描述性信息，比如说 "a police officer who works with an anti-gang unit"，而不是单纯地写 "a police officer"。b.在脚注中向读者解释为什么你要省去受访者的姓名，例如：All interviews with State Department officials were conducted with guarantees of anonymity because the officials were not authorized to disclose this information to the public. [99] Ta-Nehisi Coates, "On Charlottesville, Trump, the Confederacy, Reparations & More," interview by Amy Goodman and Juan González, *Democracy Now!*, August 15, 2017, https://www.democracynow.org/2017/8/15/full_interview_ta_nehisi_coates_on.
	简略注释	[99] Robert Alter, personal interview, October 21, 2015. [99] George Lucas, video interview, October 27, 2016. [99] Nicolas Sarkozy, telephone interview, May 5, 2013. [99] Anonymous US soldier, interview by author, February 19, 2017. [99] Discussion with senior official at Department of Homeland Security, January 7, 2010. [99] Coates, "On Charlottesville."
	参考文献	▶ 如果访谈内容已出版，或可以在网络、存档中找到，则应将其列入参考文献，方便其他研究者查询验证。如果是公众无法获取的私人通信或访谈，应该在注释中完整描述，无需写入参考文献。因此上述案例中，Coates的访谈需要包含在参考文献列表内，Alter、Sarkozy、Lucas、匿名美国士兵以及高级官员无需体现在参考文献内 Coates, Ta-Nehisi. "On Charlottesville, Trump, the Confederacy, Reparations, & More." Interview by Amy Goodman and Juan González. *Democracy Now!*, August 15, 2017. https://www.democracynow.org/2017/8/15/full_interview_ta_nehisi_coates_on.
诗歌	首见注释	[99] Elizabeth Bishop, "The Fish," in *The Complete Poems, 1927-1979* (New York: Noonday Press / Farrar, Straus and Giroux, 1983), 42–44. [99] Walt Whitman, "Song of Myself," in *Leaves of Grass* (Philadelphia: David McKay, 1891–92), sec. 51, p. 78, http://whitmanarchive.org/published/LG/1891/whole.html.
	简略注释	[99] Bishop, "The Fish," 42–44. [99] Whitman, "Song of Myself," sec. 51, p. 78.
	参考文献	Bishop, Elizabeth. "The Fish." In *The Complete Poems, 1927–1979*, 42–44. New York: Noonday Press / Farrar, Straus and Giroux, 1983.

续表

诗歌	参考文献	Walt Whitman. "Song of Myself." In *Leaves of Grass*, 29–79. Philadelphia: David McKay, 1891–92. http://whitmanarchive.org/published/LG/1891/whole.html.
戏剧剧本	首见注释	⁹⁹ Shakespeare, *Hamlet, Prince of Denmark*, 2.1.1–9. ▶ 指的是第二幕第一场，第 1 至 9 行 ▶ 如果引用了某个特定的版本，可以写作： ⁹⁹ William Shakespeare, *Hamlet, Prince of Denmark*, ed. Constance Jordan (New York: Pearson/Longman, 2005). ⁹⁹ William Shakespeare, *The Three-Text Hamlet: Parallel Texts of the First and Second Quartos and First Folio*, ed. Bernice W. Kliman and Paul Bertram (New York: AMS Press, 2003). ⁹⁹ William Shakespeare, *Hamlet, Prince of Denmark* (1600–1601; University of Virginia Library, Electronic Text Center, 1998), http://etext.virginia.edu/toc/modeng/public/MobHaml. html.
	简略注释	⁹⁹ Shakespeare, *Hamlet*, 2.1.1–9.
	参考文献	Shakespeare, William. *Hamlet, Prince of Denmark*. Edited by Constance Jordan. New York: Pearson/Longman, 2005. Shakespeare, William. *The Three-Text Hamlet: Parallel Texts of the First and Second Quartos and First Folio*. Edited by Bernice W. Kliman and Paul Bertram. New York: AMS Press, 2003. Shakespeare, William. *Hamlet, Prince of Denmark*. 1600–1601. University of Virginia Library, Electronic Text Center, 1998. http://etext.virginia.edu/toc/modeng/public/MobHaml. html.
戏剧或舞蹈演出	首见注释	⁹⁹ *Romeo and Juliet*, choreography by Krzysztof Pastor, music by Sergei Prokofiev, Joffrey Ballet, Auditorium Theatre, Chicago, October 13, 2016. ⁹⁹ *Fake*, written and directed by Eric Simonson, performed by Kate Arrington, Francis Guinan, and Alan Wilder, Steppenwolf Theatre, Chicago, October 28, 2009. ▶ 如果你关注某个人物或某个职位，就把对应的名字放在最前面。比如，你需要突出 Kate Arrington 的表演，那么可以将注释写成： ⁹⁹ Kate Arrington, performance, *Fake*, written and directed by Eric Simonson, ...
	简略注释	⁹⁹ *Romeo and Juliet*. ⁹⁹ *Fake*.
	参考文献	*Romeo and Juliet*. Choreography by Krzysztof Pastor. Music by Sergei Prokofiev. Joffrey Ballet. Auditorium Theatre, Chicago, October 13, 2016. *Fake*. Written and directed by Eric Simonson. Performed by Kate Arrington, Francis Guinan, and Alan Wilder. Steppenwolf Theatre, Chicago, October 28, 2009. ▶ 或者，如果你特别关注 Arrington 的表演： Kate Arrington, performance. *Fake*. Written and directed by Eric Simonson, ... ▶ 现场演出经常不列入参考文献，理由是读者无法亲自验证当时的内容
电视节目	首见注释	⁹⁹ *Game of Thrones*, season 4, episode 10, "The Children," June 15, 2014. ▶ 或列出更完整的出处： ⁹⁹ *Game of Thrones*, season 4, episode 10, "The Children," directed by Alex Graves, written by David Benioff and D. B. Weiss, featuring Peter Dinklage, Nikolaj Coster-Waldau, Lena Heady, Kit Harington, and Maisie Williams, aired June 15, 2014, on HBO.

电视节目	首见注释	[99] *30 Rock*, season 4, episode 1, "Season 4," directed by Don Scardino, written by Tina Fey, performed by Tina Fey, Tracy Morgan, Jane Krakowski, Jack McBrayer, Scott Adsit, Judah Friedlander, and Alec Baldwin, aired October 15, 2009 (NBC), http://www.hulu.com/30-rock. ▶ 你没看错，上例中第四季的第一集就叫"Season 4"，所以提供准确的集数会更有帮助。附上首播日期同样是为了让信息更加精准
	简略注释	[99] *Game of Thrones*, "The Children." [99] *30 Rock*, "Season 4."
	参考文献	*Game of Thrones*. Season 4, episode 10, "The Children," Directed by Alex Graves. Written by David Benioff and D. B. Weiss. Featuring Peter Dinklage, Nikolaj Coster-Waldau, Lena Heady, Kit Harington, and Maisie Williams. Aired June 15, 2014, on HBO. *30 Rock*. Season 4, episode 1, "Season 4." Directed by Don Scardino. Written by Tina Fey. Performed by Tina Fey, Tracy Morgan, Jane Krakowski, Jack McBrayer, Scott Adsit, Judah Friedlander, and Alec Baldwin. Aired October 15, 2009, on NBC. http://www.hulu.com/30-rock.
电影	首见注释	[99] *Godfather II*, directed by Francis Ford Coppola (1974; Los Angeles: Paramount Home Video, 2003), DVD. [99] *Tig*, directed by Kristina Goolsby and Ashley York (Beachside Films, 2015), Netflix. ▶ 如果你想引用个别的场景，可以仿照著作中某个章节的格式来处理，写成："Murder of Fredo," in *Godfather II* …
	简略注释	[99] *Godfather II*. [99] *Tig*.
	参考文献	*Godfather II*, DVD. Directed by Francis Ford Coppola. Performed by Al Pacino, Robert De Niro, Robert Duvall, Diane Keaton. Screenplay by Francis Ford Coppola and Mario Puzo based on novel by Mario Puzo. 1974; Paramount Home Video, 2003. *Tig*. Netflix. Directed by Kristina Goolsby and Ashley York. Written by Jennifer Arnold. Performed by Tig Notaro, Stephanie Allynne, Zach Galifianakis, Sarah Silverman. Beachside Films, 2015. ▶ 影片名、导演、电影公司和上映年份都是必要信息。如果引用的是影片的录像带，则应注明录像带的发行年份 ▶ 可以灵活处理的是：演员、制片人、编剧、剪辑、摄像等人物信息。可以根据实际写作需求决定是否列出，并按照相对于论文的重要性进行排序，把涉及的姓名写在片名和发行商之间
艺术作品	首见注释	[99] Jacopo Robusti Tintoretto, *The Birth of John the Baptist*, ca. 1550, Hermitage, St. Petersburg. ▶ 如果不清楚艺术作品产生的确切时间，可以给出一个近似时间。一幅诞生在"circa 1550"（约 1550 年）的画作，应缩写成"ca. 1550" [99] Jacopo Robusti Tintoretto, *The Birth of John the Baptist*, 1550s, in Tom Nichols, *Tintoretto: Tradition and Identity* (London: Reaktion Books, 1999), 47. ▶ 如果是复制品，还应标明它的来源 [99] Jacopo Robusti Tintoretto, *The Birth of John the Baptist* (detail), 1550s, Hermitage, St. Petersburg, http://cgfa.acropolisinc.com/t/p-tintore1.htm. ▶ 如果是在线的艺术作品，应写清网址

续表

艺术作品	简略注释	[99] Tintoretto, *Birth of John the Baptist*. ▶ 原版、复制品或线上作品的简化原则都是一样的
	参考文献	▶ 艺术作品、雕塑或照片均不列入参考文献
照片	首见注释	[99] Ansel Adams, *Monolith, the Face of Half Dome, Yosemite National Park*, 1927, Art Institute, Chicago. [99] Ansel Adams, *Monolith, the Face of Half Dome, Yosemite National Park*, 1927, Art Institute, Chicago, http://www.hctc.commnet.edu/artmuseum/anseladams/details/pdf/monlith.pdf.
	简略注释	[99] Adams, *Monolith*.
	参考文献	▶ 艺术作品、雕塑或照片均不列入参考文献
图：地图、图表、曲线图或表格	首见注释	▶ 地图、图表、曲线或表格的引用方式是直接在其下方标明来源，或在正文中说明出处；如果确有需要，也可以将来源信息写在注释中 Source: Ken Menkhaus, "Governance without Government in Somalia: Spoilers, State Building, and the Politics of Coping," *International Security* 31 (Winter 2006/7): 79, fig. 1. Source: "Presidential Election Results: Donald J. Trump Wins" (interactive map), *New York Times*, August 9, 2017, https://www.nytimes.com/elections/results/president. [99] Garrett Dash Nelson and Alasdair Rae, *An Economic Geography of the United States: From Commutes to Megaregions, Plos One*, accessed August 29, 2017, https://doi.org/10.1371/journal.pone.0166083.g001. [99] G. M. Wheeler, *Topographical Map of the Yosemite Valley and Vicinity*, preliminary edition, 1883, 43 × 55cm, 1:42,240 scale, David Rumsey Historical Map Collection, http://www.davidrumsey.com/luna/servlet/detail/RUMSEY~8~1~405~40014:Topographical-Map-Of-The-Yosemite-V. Source: Google Maps, map of 1427 E. 60th St., Chicago, IL 60637, accessed May 5, 2017, http://maps.google.com/. ▶ 引用地图的时候，应标注制图师（如果知道相关信息的话）、地图的标题或描述、比例尺和尺寸、出版详情或出版地等信息 ▶ 如果线上的地图没有绘制日期，则应注明访问日期或修订日期
	简略注释	▶ 为所有的图使用完整注释
	参考文献	Menkhaus, Ken. "Governance without Government in Somalia: Spoilers, State Building, and the Politics of Coping." *International Security* 31 (Winter 2006/7): 74–106. ▶ 不需要在参考文献中列出地图、图表、曲线或表格。但如果它们取自书籍、文章等出版物，则应写入参考文献。谷歌地图无需出现在参考文献，其他来源需要 "Presidential Election Results: Donald J. Trump Wins." *New York Times*, August 9, 2017. https://www.nytimes.com/elections/results/president.
音乐唱片	首见注释	[99] Robert Johnson, "Cross Road Blues," 1937, *Robert Johnson: King of the Delta Blues Singers*, Columbia Records 1654, 1961. [99] Samuel Barber, "Cello Sonata, for cello and piano, op. 6," *Barber: Adagio for Strings, Violin Concerto, Orchestral and Chamber Works*, compact disc 2, St. Louis Symphony, cond. Leonard Slatkin, Alan Stepansky (cello), Israela Margalit (piano), EMI Classics 74287, 2001.

音乐唱片	首见注释	⁹⁹ Jimi Hendrix, "Purple Haze," 1969, *Woodstock: Three Days of Peace and Music*, compact disc 4 of 4, Atlantic/Wea, 1994. ⁹⁹ Vladimir Horowitz, Hungarian Rhapsody no. 2, by Franz Liszt, recorded live at Carnegie Hall, February 25, 1953, MP3 file (241 Kbps). ▶ Kbps 代表 kilobits per second（码率，千位每秒），MP3 是一种常见的音频的压缩格式。如果 MP3 文件是从光盘（compact disc）或黑胶唱片（LP）上复制的版本，你可以写出最初的音源： ⁹⁹ Vladimir Horowitz, Hungarian Rhapsody no. 2, by Franz Liszt, recorded live at Carnegie Hall, February 25, 1953, compact disc, RCA Victor 60523-2-RG, 1991. ⁹⁹ Beyoncé, "Formation," TIDAL, track 11 on *Lemonade*, Parkwood Entertainment and Columbia Records, 2016. ▶ 还应写清音频流的文件格式或服务提供商
	简略注释	⁹⁹ Johnson, "Cross Road Blues." ⁹⁹ Barber, "Cello Sonata, op. 6." ⁹⁹ Hendrix, "Purple Haze." ⁹⁹ Horowitz, Hungarian Rhapsody no. 2. ⁹⁹ Beyoncé, "Formation."
	参考文献	Johnson, Robert. "Cross Road Blues." 1937. *Robert Johnson: King of the Delta Blues Singers*. Columbia Records 1654, 1961. Barber, Samuel. "Cello Sonata, for cello and piano, op. 6." *Barber: Adagio for Strings, Violin Concerto, Orchestral and Chamber Works*. Compact disc 2. St. Louis Symphony. Cond. Leonard Slatkin, Alan Stepansky (cello), Israela Margalit (piano). EMI Classics 74287, 2001. Hendrix, Jimi. "Purple Haze," 1969. *Woodstock: Three Days of Peace and Music*. Compact disc 4. Atlantic/Wea, 1994. Horowitz, Vladimir. Hungarian Rhapsody no. 2, by Franz Liszt. Recorded live at Carnegie Hall, February 25, 1953. MP3 file (241 Kbps). Beyoncé. "Formation." TIDAL. Track 11 on *Lemonade*. Parkwood Entertainment and Columbia Records, 2016.
朗读、讲座或有声书的录音	首见注释	⁹⁹ Franklin D. Roosevelt, "Fireside Chat: Outlining New Deal Program," May 7, 1933, Recorded Speeches and Utterances of Franklin D. Roosevelt, FDR Presidential Library and Museum, MP3 audio (23:00), http://www.fdrlibrary.marist.edu/archives/collections/utterancesfdr. html. ⁹⁹ George Eliot, *Middlemarch*, performed by Juliet Stevenson, Naxos Audiobooks, released on Audible .com, March 18, 2011. ▶ 朗读、书籍或讲座录音的引用方法与音乐唱片的引用方法类似
	简略注释	⁹⁹ Roosevelt, "Fireside Chat." ⁹⁹ Eliot, *Middlemarch*.
	参考文献	Roosevelt, Franklin D. "Fireside Chat: Outlining New Deal Program." May 7, 1933. Recorded Speeches and Utterances of Franklin D. Roosevelt, FDR Presidential Library and Museum. MP3 audio (23:00). http://www.fdrlibrary.marist.edu/archives/collections/ utterancesfdr.html. Eliot, George. *Middlemarch*. Performed by Juliet Stevenson. Naxos Audiobooks. Released on Audible .com, March 18, 2011.

音乐录像带、相关的评论	首见注释	⁹⁹ Arcade Fire, "Everything Now," music video, directed by the Sacred Egg, released June 1, 2017, https://youtu.be/zC30BYR3CUk. ⁹⁹ Rihanna, Kanye West, and Paul McCartney, "FourFiveSeconds," music video, directed by Inez & Vinoodh, produced by Stephanie Bargas and Jeff Lepine, February 3, 2015, http://vevo.ly/3aZlDK. ⁹⁹ Bruno Mars, "That's What I Like," music video, directed by Bruno Mars and Jonathan Lia, March 1, 2017, https://youtu.be/PMivT7MJ41M. ⁹⁹ Minimanaloto, March 3, 2017, comment on Mars, "That's What I Like."
	简略注释	⁹⁹ Arcade Fire, "Everything Now." ⁹⁹ Rihanna, Kanye West, and Paul McCartney, "FourFiveSeconds." ⁹⁹ Mars, "That's What I Like." ⁹⁹ Minimanaloto, comment on Mars, "That's What I Like."
	参考文献	Arcade Fire. "Everything Now." Music video. Directed by the Sacred Egg, released June 1, 2017. https://youtu.be/zC30BYR3CUk. Rihanna, Kanye West, and Paul McCartney. "FourFiveSeconds." Music video. Directed by Inez & Vinoodh. Produced by Stephanie Bargas and Jeff Lepine, released February 3, 2015, http://vevo.ly/3aZlDK. Mars, Bruno. "That's What I Like." Music video. Directed by Bruno Mars and Jonathan Lia, released March 1, 2017. https://youtu.be/PMivT7MJ41M. ▶ 观众对音乐录像带的评论不放入参考文献列表中引用
活页乐谱	首见注释	⁹⁹ Johann Sebastian Bach, "Toccata and Fugue in D Minor," 1708, BWV 565, arranged by Ferruccio Benvenuto Busoni for solo piano (New York: G. Schirmer LB1629, 1942).
	简略注释	⁹⁹ Bach, "Toccata and Fugue in D Minor."
	参考文献	Bach, Johann Sebastian. "Toccata and Fugue in D Minor." 1708. BWV 565. Arranged by Ferruccio Benvenuto Busoni for solo piano. New York: G. Schirmer LB1629, 1942. ▶ 这首乐曲写于 1708 年，具有 Bach 作品的标准分类编码 BWV 565。这个特别的编曲是由 G. Schirmer 在 1942 年发行的，Schirmer 对它的编号是 LB1629
唱片内页文字说明	首见注释	⁹⁹ Steven Reich, liner notes for *Different Trains*, Elektra/Nonesuch 9 79176-2, 1988.
	简略注释	⁹⁹ Reich, liner notes. ▶ 或 ⁹⁹ Reich, liner notes, *Different Trains*.
	参考文献	Reich, Steven. Liner notes for *Different Trains*. Elektra/Nonesuch 9 79176-2, 1988.
广告	首见注释	⁹⁹ *Letters from Iwo Jima advertisement, New York Times,* February 6, 2007, B4. ▶ 鉴于报纸的版本众多、分页方式不同，页码信息是可以省略的 ⁹⁹ Vitamin Water, "Drink outside the Lines" advertisement, *Rolling Stone*, June 15, 2017, 17. ⁹⁹ Tab cola, "Be a Mindsticker," television advertisement, ca. late 1960s, http://www.dailymotion.com/video/x2s3qi_1960s-tab-commercial-be-a-mindstick_ads.
	简略注释	⁹⁹ *Letters from Iwo Jima advertisement.* ⁹⁹ Vitamin Water, "Drink outside the Lines" advertisement. ⁹⁹ Tab advertisement.

续表

广告	参考文献	▶ 极少将广告的来源列入参考文献，除非它对你的论文极其重要 *Letters from Iwo Jima advertisement. New York Times,* February 6, 2007. Vitamin Water. "Drink outside the Lines" advertisement. *Rolling Stone*, June 15, 2017, 17. Tab cola. "Be a Mindsticker." Television advertisement, ca. late 1960s. http://www.dailymotion.com/video/x2s3qi_1960s-tab-commercial-be-a-mindstick_ads.
政府文献	首见注释	[99] Senate Committee on Armed Services, *Hearings on S. 758, A Bill to Promote the National Security by Providing for a National Defense Establishment,* 80th Cong., 1st sess., 1947, S. Rep. 239, 13. ▶ "S. Rep. 239, 13" 代表 239 号报告的第 13 页 [99] *Financial Services and General Government Appropriations Act, 2008,* HR 2829, 110th Cong., 1st sess., Congressional Record 153 (June 28, 2007): H7347. [99] *American Health Care Act, 2017,* HR 1628, 115th Cong., 1st sess., June 8, 2017, Congress.Gov, https://www.congress.gov/bill/115th-congress/house-bill/1628/text. [99] Department of the Treasury, Office of Foreign Assets Control (OFAC), *Sanctions Actions Pursuant to Executive Order 13581, Federal Register* 82, no. 166 (August 29, 2017): 41019, https://www.gpo.gov/fdsys/pkg/FR-2017-08-29/pdf/2017-18289.pdf. [99] US Department of State, Daily Press Briefing, August 9, 2017, https://www.state.gov/r/pa/prs/dpb/2017/08/273288.htm.
	简略注释	[99] Senate, *Hearings on S. 758,* 13. [99] *American Health Care Act,* HR 1628. [99] OFAC, *Sanctions Actions.* [99] US State Department, Daily Press Briefing, August 9, 2017.
	参考文献	US Congress. Senate. Committee on Armed Services. *Hearings on S. 758, Bill to Promote the National Security by Providing for a National Defense Establishment.* 80th Cong., 1st sess., 1947. S. Rep. 239. US Congress. House. *Congressional Record.* 110th Cong., 1st sess. June 28, 2007. Vol. 153, no. 106. H7347. *American Health Care Act, 2017.* HR 1628, 115th Cong., 1st sess. June 8, 2017. Congress.Gov, https://www.congress.gov/bill/115th-congress/house-bill/1628/text. Department of the Treasury, Office of Foreign Assets Control. *Sanctions Actions Pursuant to Executive Order 13581. Federal Register* 82, no. 166 (August 29, 2017): 41019. https://www.gpo.gov/fdsys/pkg/FR-2017-08-29/pdf/2017-18289.pdf. US Department of State. Daily press briefing. August 9, 2017. https://www.state.gov/r/pa/prs/dpb/2017/08/273288.htm.
网站、网页	首见注释	[99] "News and Advocacy," American Historical Association (website), accessed August 1, 2017, https://www.historians.org/news-and-advocacy. ▶ 许多网站没有明确的作者，建议列出赞助该网站的公司或组织 ▶ 如果从标题不能明显看出这是一个网站或网页，请补充 "website" 或 "webpage" 的字样 [99] Digital History (website), ed. Steven Mintz, accessed August 5, 2017, http://www.digitalhistory.uh.edu/. [99] Yale University, History Department home page, accessed August 5, 2017, http://www.yale.edu/history/. ▶ 如果明确看出来源是网站的主页，可以省去 "home page" 等词

网站、网页	首见注释	[99] Charles Lipson, "News and Commentary: US and the World," accessed August 29, 2017, http://www.charleslipson.com/News-links.htm. ▶ 建议注明发布日期或修订日期。如果找不到这类信息，则注明访问日期
	简略注释	[99] "News and Advocacy." [99] Digital History. [99] Yale History Department. [99] Lipson, "News and Commentary."
	参考文献	American Historical Association (website). "News and Advocacy." Accessed August 1, 2017. https://www.historians.org/news-and-advocacy. Digital History (website). Edited by Steven Mintz. Accessed August 5, 2017. http://www.digitalhistory.uh.edu/. Yale University. History Department home page. Accessed August 5, 2017. http://www.yale.edu/history/. Lipson, Charles. "News and Commentary: US and the World." Accessed August 29, 2017. http://www.charleslipson.com/News-links.htm. ▶ 网站的引用都放在注释里就可以了。如果你把它们写入参考文献，需要列出网站的所有者、作者或赞助商
博客文章或评论	首见注释	[99] Greg Weeks, "Venezuela's Core Support," *Two Weeks Notice: A Latin American Politics Blog*, August 2, 2017, http://weeksnotice.blogspot.com/2017/08/venezuelas-core-support.html. ▶ 如果文章没有标题，则可写作：Greg Weeks, untitled post, *Two Weeks Notice: A Latin American Politics Blog*, ... [99] Ana Arana, "The Deep Magic of Mexico," *The Big Roundtable* (blog), December 8, 2016, https://thebigroundtable.com/the-deep-magic-of-mexico-2bf1b341d1a1. [99] Stefanie Fletcher, December 8, 2016 (11:30 a.m.), comment on Arana, "Deep Magic." ▶ Stefanie Fletcher 发表了不止一条评论，所以引用时添加了具体的时间信息 ▶ 评论需要依托原帖进行引用
	简略注释	[99] Weeks, "Venezuela's Core Support." [99] Arana, "Deep Magic." [99] Fletcher, comment on Arana, "Deep Magic."
	参考文献	▶ 博客文章通常不列入参考文献。如果文章本身特别重要或是对你的论文很关键，可以将参考文献写作： Weeks, Greg. "Venezuela's Core Support." *Two Weeks Notice: A Latin American Politics Blog*, August 2, 2017. http://weeksnotice.blogspot.com/2017/08/venezuelas-core-support.html. Arana, Ana. "The Deep Magic of Mexico." *The Big Roundtable* (blog), December 8, 2016. https://thebigroundtable.com/the-deep-magic-of-mexico-2bf1b341d1a1.
视频	首见注释	[99] *Duck and Cover*, Federal Civil Defense Administration / Archer Productions, 1951, video file, posted August 19, 2008, http://en.wikipedia.org/wiki/File: DuckandC1951.ogg. [99] "Martin Shkreli on Three Counts of Security Fraud," NBC video news clip, August 4, 2017, http://www.nbcnews.com/video/jury-convicts-martin-shkreli-on-three-counts-of-securities-fraud-1017719875664.

续表

视频	首见注释	⁹⁹ Franchesca Ramsey, "Turn Your Phone!" Franchesca Ramsey video blog, accessed August 5, 2017, 3:07, http://www.franchesca.net/videos-photos/. ⁹⁹ Tricia Wang, "The Human Insights Missing from Big Data," filmed September, 2016, TEDxCambridge, TED video, 16:12, https://www.ted.com/talks/tricia_wang_the_human_insights_missing_from_big_data.
	简略注释	⁹⁹ *Duck and Cover.* ⁹⁹ "Jury Convicts Martin Shkreli," August 4, 2017. ⁹⁹ Ramsey, "Turn Your Phone!" ⁹⁹ Wang, "Human Insights."
	参考文献	*Duck and Cover.* Federal Civil Defense Administration / Archer Productions, 1951. Video file. Posted August 19, 2008. http://en.wikipedia.org/wiki/File: DuckandC 1951.ogg. "Jury Convicts Martin Shkreli on Three Counts of Security Fraud." NBC video news clip, August 4, 2017. http://www.nbcnews.com/video/jury-convicts-martin-shkreli-on-three-counts- of-securities-fraud-1017719875664. Ramsey, Franchesca. "Turn Your Phone!" Franchesca Ramsey video blog. Accessed August 5, 2017. 3:07. http://www.franchesca.net/videos-photos/. Wang, Tricia. "The Human Insights Missing from Big Data." Filmed September, 2016, at TEDxCambridge. TED video, 16:12. https://www.ted.com/talks/tricia_wang_the_human_insights_missing_from_big_data.
多媒体应用（电子游戏以及其他独立软件）	首见注释	⁹⁹ *Madden NFL 18* (EA Sports, November 2017), PlayStation 4, Xbox One. ⁹⁹ *Geocaching*, v. 5.6.1 (Seattle: Groundspeak, 2017), Android, iOS 9.0 or later. ▶ 尽量将版本号、发布日期、允许软件所需的设备和系统等信息写完整
	简略注释	⁹⁹ *Madden NFL 18.* ⁹⁹ *Geocaching.*
	参考文献	EA Sports. *Madden NFL 18*. November 2017, PlayStation 4, Xbox One. Groundspeak. *Geocaching*, v. 5.6.1. Seattle: Groundspeak, 2017. Android, iOS 9.0 or later.
播客	首见注释	⁹⁹ Brian Reed, host, "Has Anybody Called You?" *S-Town*, podcast, produced by Brian Reed and Julie Snyder, Serial Productions, March 28, 2017, https://stownpodcast.org/chapter/2. ⁹⁹ Ellen DeGeneres, "Ellen's Parenting Advice," *The Ellen Show Podcast*, Telepictures Productions, video podcast, June 26, 2015, iTunes. ▶ 若不能找到播客节目的发布时间，需提供访问日期
	简略注释	⁹⁹ Reed, "Chapter 2." ⁹⁹ DeGeneres, "Ellen's Parenting Advice."
	参考文献	Reed, Brian, host. "Has Anybody Called You?" *S-Town*, podcast. Brian Reed and Julie Snyder, Serial Productions. March 28, 2017. https://stownpodcast.org/chapter/2 DeGeneres, Ellen. "Ellen's Parenting Advice." *The Ellen Show Podcast*. Telepictures Productions. Video podcast. June 26, 2015. iTunes.
社交媒体 (Facebook, Instagram, Twitter)	首见注释	⁹⁹ Neil deGrasse Tyson (@neiltyson), "Always seemed to me that millipedes have more legs than are necessary," Twitter, July 7, 2017, 1:25 p.m., twitter.com/neiltyson/status/ 883421859072458752.

续表

社交媒体 (Facebook, Instagram, Twitter)	首见注释	⁹⁹ Smithsonian National Museum of African American History and Culture, "For African Americans, the pleasure of swimming at a public pool or beach was not always a right," Facebook, July 28, 2017, 1:50 p.m., https://www.facebook.com/NMAAHC/. ⁹⁹ Bruce Jackson, "Kennywood closed and paved over the pool rather than integrate," July 30, 2017, 11:49 a.m., comment on Smithsonian, "For African Americans." ⁹⁹ Cory Richards (@coryrichards), "A quiet if not solemn day on Everest," Instagram photo, May 24, 2017, https://www.instagram.com/p/BUekBZLgDOW/?hl=en. ▶ 公开的社交媒体内容，包括对他人帖子的评论，应以上述方式引用。非公开内容，例如私信，需要以"私人通信"的形式来引用 ▶ 评论需要依托原帖进行引用 ▶ 如果原帖没有标题，可以使用正文（最多截取160字符） ▶ 若不知道作者真实身份，可以仅提供网名
	简略注释	⁹⁹ Tyson, "Always seemed to me." ⁹⁹ Smithsonian, "For African Americans." ⁹⁹ Jackson, comment on Smithsonian, "For African Americans."
	参考文献	⁹⁹ Richards, "A quiet if not solemn day." ▶ 社交网站上的个人评论不需要写入参考文献。出于写作的实际需求，可以引用某个用户的页面 Richards, Corey. Instagram page. https://www.instagram.com/p/BUekBZLgDOW/?hl=en.
邮件、短信 （私信）	首见注释	⁹⁹ Jim Abrams, "Fw: a Favor," email from Jim Abrams to Mayor Rahm Emanuel, April 28, 2015, https://assets.documentcloud.org/documents/3516674/Abrams-amp-Bank-Phoenix-Electric.pdf. ▶ 引用邮件内容时，无需提供访问日期 ▶ 如果邮件已存储归档，需提供网址 ⁹⁹ Kathy Leis, email message to author, May 5, 2017. ⁹⁹ Kathy Leis, text message to author, May 5, 2017. ⁹⁹ Alison Viup, Facebook direct message to author, July 20, 2016. ▶ 如果某条电子信息的时间很重要，或需要与其他信息区别开来，则可注明详细时间。比如： ⁹⁹ Michael Lipson, text message to Jonathan Lipson, May 5, 2017, 11:23 a.m. ▶ 若准备引用某条电子信息，应注意保存副本
	简略注释	⁹⁹ Jim Abrams to Rahm Emanuel, April 28, 2015. ⁹⁹ Kathy Leis, email message to author, May 3, 2017. ▶ 可以尝试尽量写得简洁。如果只涉及一位名叫 Leis 的发信人，则可只写姓氏。但如果引用的信息来自 Alan Leis、Kathy Leis、Elizabeth Leis、Julia Leis 等人，那么在简略注释里也要写清全名 ⁹⁹ Michael Lipson, instant message to Jonathan Lipson, May 5, 2017. ⁹⁹ Michael Lipson, text message to Jonathan Lipson, May 5, 2017, 11:23 a.m.
	参考文献	▶ 个人邮件、短信和其他即时信息以及发送给邮件列表的邮件不必列入参考文献，除非它们可以被第三方检索到 Abrams, Jim. "Fw: a Favor." Email from Jim Abrams to Mayor Rahm Emanuel, April 28, 2015. https://assets.documentcloud.org/documents/3516674/Abrams-amp-Bank-Phoenix-Electric.pdf.

续表

电子论坛、邮件列表	首见注释	⁹⁹ Quinn Ngo, answer to "What Are Some Unknown Truths about College Life?," Quora, July 14, 2017, https://www.quora.com/What-are-some-unknown-truths-about-college-life. ⁹⁹ Mablegrable, "Cozy spot for a morning coffee," Reddit photo, August 4, 2017, https://i.redd.it/okqb36mkbndz.jpg. ⁹⁹ VeganSweets, "There's nothing like nature and some coffee/tea," August 4, 2017, 1:28 p.m., comment on Mablegrable, "Cozy Spot." ⁹⁹ Mark Aaron Polger, "CALL: Call for Proposals," American Library Association mailing list archives, ACRL Campaign for Libraries list, May 10, 2017, http://lists.ala.org/sympa/arc/academicpr/2017-05/msg00005.html. ▶ 如果引用的是私人邮件列表或私人讨论区上发表的信息，应视作私人通信，请仿照邮件、短信的引用方式处理
	简略注释	⁹⁹ Ngo, Quora, July 14, 2017. ⁹⁹ VeganSweets, comment on Mablegrable, "Cozy Spot." ⁹⁹ Polger, ALA mailing list, May 10, 2017.
	参考文献	▶ 个人评论类的发言无需列入参考文献。但如果你在文中大量引用某部分内容，可以考虑在参考文献中列出整个主题或整个论坛 Quora. "What Are Some Unknown Truths about College Life?" July 14, 2017. https://www.quora.com/What-are-some-unknown-truths-about-college-life.

表6.3　芝加哥引用格式：如何引用注释和表格

写法	代表的含义
106	第 106 页
106n	第 106 页的唯一一个注释
107n32	第 107 页的第 32 条注释（当前页面有多条注释）
89, table 6.2	第 89 页上的表 6.2（曲线、图形的标注方式与之类似）

表6.4　芝加哥引用格式：常见的缩写

全称	缩写	全称	缩写	全称	缩写	全称	缩写
and others	et al.	edited by	ed.	notes	nn.	part	pt.
appendix	app.	edition	ed.	number	no.	Pseudonym	pseud.
book	bk.	editor(s)	ed(s).	opus	op.	translator	trans.
chapter	chap.	especially	esp.	opuses	opp.	versus	vs.
compare	cf.	figure	fig.	page	p.	volume	vol.
document	doc.	note	n.	pages	pp.		

注：缩写都用小写字母，后面紧跟英文句号。大部分缩写的复数形式为直接加 s，特例是 note (n.→nn.)、opus (op.→opp.)、page (p.→pp.) 和 translator（复数为 trans. 不变）。（另外请注意：当缩写 n. 或 nn. 紧跟在页码数字后面时，英文句号应删掉。）

引用诗歌时，不要把 line 或 lines 用缩写取代，因为小写字母 l 容易同数字 1 混淆。

6.2 芝加哥引用格式的常见问题

（1）芝加哥引用格式和杜拉宾式有什么区别吗？

没有。二者曾有细微差别（例如是否针对在线资料标注访问日期），但目前看来，二者基本一致。

（2）对于在线资料，是否需要列出访问日期？

一般来说不需要，但前提是资料来源上能找到正式的出版、发布日期或最近的修改日期。如果无法列出上述日期，那么访问日期是必要的——它至少能帮助读者确定该材料在网站上的可用时间。此外，即使能列出发布或修改日期，一些老师仍会要求学生注明访问日期，建议你事先询问具体要求。

（3）为什么要在一些出版社的后面列出它所在的州，而另一些却没有写明？

The Chicago Manual of Style 的建议是：除非出版社位于大型、著名城市，否则要写出州名。为避免混淆，对于哈佛大学出版社、麻省理工学院出版社，使用 Cambridge, MA（马萨诸塞州的剑桥）来标记；对于英国的剑桥大学出版社，则直接用 Cambridge。同时，如果州名已经包含在出版社的名称中，也可不再重复列出，比如：Ann Arbor: University of Michigan Press。

（4）如何表示即将出版的图书？

在出版年份那里标注"forthcoming"一词。以参考文献的写法为例：
Godot, Shlomo. *Still Waiting*. London: Verso, forthcoming.

（5）出版时间和地点均为不详，该怎么办？

与刚才的思路相似，在时间或地点那里标注"n.d."（no date）或"n.p."（no place）。例如：Montreal, QC: McGill-Queen's University Press, n.d.

（6）根据引用格式，如果被引图书的标题以问号结尾，后面是否还要使用逗号或句号？

如果需要逗号，请照常添加。但不要在其后面再使用英文句号。

（7）注释是单倍行距还是双倍行距？参考文献呢？

脚注、尾注的行距与正文相同即可。

对于参考文献，我个人认为：同一条文献内部用单倍行距，不同条目之间用双倍行距，这样最有利于阅读。这个问题最好咨询你所在的院系或合作的出版社——他们可能要求全篇都用双倍行距来排版。

（8）我读了马克·吐温的作品，引用时该用哪个名字，Twain 还是 Samuel Clemens?

像 Mark Twain 或 Mother Teresa 这种家喻户晓的笔名是允许直接使用的，无需特殊说明。

如果你想补充他们的本名，也很容易达成。把人们熟知的名字放在前面（本名或假名均可），然后把另一个名字放入方括号，写在后面。请看几个案例：

George Eliot [Mary Ann Evans]
Isak Dinesen [Karen Christence Dinesen, Baroness Blixen-Finecke]
Le Corbusier [Charles-Edouard Jeanneret]
Benjamin Disraeli [Lord Beaconsfield]
Lord Palmerston [Henry John Temple]
Krusty the Clown [Herschel S. Krustofski]

如果你希望将假名写入参考文献的条目，可以这样做：
Aleichem, Sholom [Solomon Rabinovitz]. *Fiddler on the Roof* ...

7 MLA 格式：人文学科

7.1 MLA 格式的用法

美国现代语言学会（MLA）制定的引用格式在人文学科中有广泛应用。该格式采用（Strier 125）风格的文内夹注，而不使用脚注或尾注。相应条目的完整信息列入文后的参考文献（Works Cited）；与其他引用格式类似，每条文献都包含三类信息：作者、标题和出版信息。以 Amitav Ghosh 的一本著作为例，它在 Works Cited 部分的完整形式为：

Ghosh, Amitav. *The Great Derangement: Climate Change and the Unthinkable.* U of Chicago P, 2016.

2016 年的第 8 版 *MLA Handbook* 与之前的版本相比，有一个重大变化。这次的官方手册为各种类型、格式的被引著作创立了一条总体原则。在过去，MLA 格式要求作者标示出被引著作的媒介（print/web/email/television 等），同时提供出版信息。但由于媒介种类不断增加，不同格式之间也能轻易转换，MLA 格式现在的理念是：取消各种死板的规则，让作者能够灵活引用不同媒介的内容。

文内夹注的方式非常简便，可以在句末插入（Ghosh）来引用该作者的整本书，也可以写成（Ghosh 12）指向第 12 页。如果文章参考了同一作者的不同作品，要让读者知道每一处对应哪本作品。一种做法是在行文时加以说明，另一种是把简短的标题写到圆括号里：（Ghosh, *Great Derangement* 12）。

在此基础上，还能继续简化——MLA 格式始终追求简洁。只要能看出引文来源，括号中的作者、标题也可以省去不写，例如：

As Ghosh notes, climate change is rarely a topic in fiction classified as "literary" (7–9).

如果不涉及具体页码，并且正文明确给出了作者和标题，整个文内夹注的括号都可以去掉。请看例句：

Gibbon's *Decline and Fall of the Roman Empire* established new standards of documentary evidence for historians.

上句本身已经能辨别作者和标题两大要素，无需额外的文内夹注。这是 MLA 格式的简洁性所决定的。但是，Gibbon 的作品仍要在参考文献中出现。

MLA 格式还允许你将多位作者的文献放进一个括号里：（Bevington 17; Bloom 75; Vendler 51），文献之间用分号隔开。

假设 Ghosh 的著作共有三卷，若引用第三卷的第 17 页，则应标注（Ghosh 3: 17）。如果希望跟 Ghosh 的其他作品区分开，则可添加标题，写成：（Ghosh, *Great Derangement* 3: 17）。如果是引用整卷书，不涉及具体某页，可以写：（Ghosh, vol. 3）或者（Ghosh, *Great Derangement*, vol. 3）。既然是以简洁著称，为何还要加上"vol."呢？一旦不标注"vol."，读者可能会误认为此处引用的是第 3 页，还会以为该作品只有一卷。

如果几位作者的姓氏相同，引注时增加名字的首字母加以区分就可以了：（C. Brontë, *Jane Eyre*），（E. Brontë, *Wuthering Heights*）。同样地，完整的作者、著作信息要列入文后的参考文献（Works Cited）。

像《简·爱》（*Jane Eyre*）这种作品或许已有无数个版本，读者可能想对照自己手里的版本来查看你引用的段落。针对这种情况，MLA 格式建议你——也就是论文的作者——在页码后面补充一些信息，方便读者查找，比如说提醒读者这段话出自第 1 章。

还是以 *Jane Eyre* 为例，假设你引用了第 1 章的某段话，它出现在你手中版本的第 7 页。那么写完页码后，你可以加一个分号，在后面列出章节信息，chapter 要用小写字母缩写：（C. Brontë, *Jane Eyre* 7; ch. 1）。

如需引用莎士比亚的作品，可以不按页码，而是按章节、场景和原文所在的行数来引用，这几项信息之间用英文句号隔开（*Romeo and Juliet* 1.3.12–15）。其他带有行数、小节标记的诗歌也可仿照上述模式进行引用；只写行数时，要用完整的单词 line 或 lines，例如（lines 12–15），不要用缩写。

网上的文件可能没有页码。如果能找到章节或段落编号，可在引用时使用，方法是先写作者名字，逗号隔开，再注明章节或段落：（Padgett, sec. 9.7）或者（Snidal, pars. 12–18）。如果网络文件为 PDF 格式，意味着所有访问者都能阅读相同的版式，所以你可以为该文件编上页码：（Wang 14）。如果确实无法给页面编号，只写作者的名字也可以。

文内夹注一般写在句末，括号后面是句子的标点。比如：

A full discussion of these issues appears in *Miss Thistlebottom's Hobgoblins* (Bernstein).

MLA 格式允许作者使用脚注或尾注，但只能用于发表评论，不得用于引用文献。如果你想在注释内部引用一些资料，按照这里讲到的文内夹注规则处理即可。

在"简洁"原则的指导下，出版商信息也可以缩写：Princeton University Press 要写成 Princeton UP，the University of Chicago Press 则写作 U of Chicago P。同理，大部分的月份名称需要以缩写形式列出。

用 MLA 风格引用电子资源也可在一定程度上简写。网址中的"http://"或"https://"无需出现（DOI 编码是例外）。虽然建议在没有出版、发布日期或最近修改日期的时候，注明访问日期，但也不是必需的。

我将在下面的表格中展示详细的引用格式，并提供例证。MLA 格式多用于人文学科，可能需要引用戏剧、诗歌和电影作品等，表格中都有列出。如需查看更多例证或引用不常见的文献类型，可以查询官方手册。

我按照字母顺序将各类条目整理成表格（表 7.1、表 7.2），同时附上页码信息，以便你迅速定位相关内容。在本章的最后，我还总结了一些 MLA 格式的常用的缩写。

表7.1 MLA格式索引表

文献类型	在本书中页码	文献类型	在本书中页码
摘要（abstract）	82	博客（blog）	
广告（advertisement）	89	评论（comment）	90
		文章、帖文（post）	90
档案材料（archival material）	83–84	图表（chart）	87
艺术作品（artwork）	87	舞蹈演出（dance performance）	86
有声书（audiobook）	88–89	词典（dictionary）	84
《圣经》（Bible）	84	私信（direct message）	91
图书（book）		学位论文（dissertation）	83
匿名作者（anonymous author）	81	电子论坛、邮件列表（electronic forum or mailing list）	91
主编图书中的一章（chapter in edited book）	82		
电子图书（e-book）	81	邮件（email）	91
主编图书（edited）	80		
多位作者（multiple authors）	80	百科全书（encyclopedia）	84
多个版本（multiple editions）	80		
多卷本著作（multivolume work）	81	脸书（Facebook）	91
无作者（no author）	81	图（figure）	87
单一作者（one author）	79		
在线图书（online）	81	电影（film）	86–87
早期版本的重印版（reprint）	82	政府文献（government document）	89
多本图书，同一作者（several by same author）	80	曲线图（graph）	87
多卷本中的单卷作品（single volume in a multivolume work）	81	照片墙（Instagram）	91
译著（translated volume）	82	访谈（interview）	85

续表

文献类型	在本书中页码	文献类型	在本书中页码
期刊论文（journal article）	82	播客（podcast）	90–91
非英语（foreign language）	83	诗歌（poem）	85
《古兰经》（Koran）	84	书评（review）	83
杂志的文章（magazine article）	83	社交媒体（social media）	91
手稿集（manuscript collection）	83–84	演讲、报告或讲座（speech, lecture, or talk）	85
地图（map）	87	表格（table）	87
多媒体应用（multimedia app）	90	电视节目（television program）	86
音乐（music）		短信（text message）	91
唱片内页文字说明（liner notes）	88	毕业论文（thesis）	83
音乐唱片（recording）	88	推特（Twitter）	91
活页乐谱（sheet music）	88		
音乐录像带（video）	88		
报纸的文章（newspaper article）	83	视频（video）	90
未发表的论文（paper, unpublished）	83	电子游戏（video game）	90
私人通信（personal communication）	85	网站、网页（website or page）	89–90
照片（photograph）	87		
戏剧（play）			
戏剧演出（performance）	86		
戏剧剧本（text）	86		

表7.2　MLA格式：参考文献和文内夹注

图书，单一作者	参考文献	Lipson, Charles. *How to Write a BA Thesis: A Practical Guide from Your First Ideas to Your Finished Paper*. U of Chicago P, 2005. Collins, Christopher. *Neopoetics: The Evolution of the Literate Imagination*. Columbia UP, 2016. Fitzpatrick, Meghan. *Invisible Scars: Mental Trauma and the Korean War*. U of British Columbia P, 2017. ▶ 如果是1900年以后出版的作品，MLA格式无需标注出版商的地理位置。如果是1900年以前的作品，不写出版商，要写城市的名字，例如： Barcelona, 1870
	文内夹注	(Lipson 80–82)或(80–82) ▶ 指的是第80至82页 ▶ 如果想区分同一作者的不同作品，引注的形式为： (Lipson, *How* 22–23) ▶ MLA格式缩写标题的原则是：保留标题的第一个名词以及它前面的所有形容词。如果标题的开头不含名词短语，若能将不同的书目区分开来，可以只保留标题的第一个词 　(Collins 26)或(26) 　(Fitzpatrick 136)或(136)

多本图书，同一作者	参考文献	Talbot, Ian. *A History of Modern South Asia: Politics, States, Diasporas.* Yale UP, 2016. ---, editor. *The Independence of India and Pakistan: New Approaches and Reflections.* Oxford UP, 2013. ---. *Pakistan: A Modern History.* Palgrave Macmillan, 2010. ▶ 重复的作者名用三条连字符代替，后面紧跟英文句号。如果该作者以编者或译者的身份再次出现，连字符后面则应使用逗号
	文内夹注	(Talbot, *History* 34; Talbot, *Independence* 3–5; Talbot, *Pakistan* 456)
图书，多位作者	参考文献	Young, Harvey, and Queen Meccasia Zabriskie. *Black Theater Is Black Life: An Oral History of Chicago Theater and Dance, 1970–2010.* Northwestern UP, 2013. Fischer, John Martin, and Benjamin Mitchell-Yellin. *Near-Death Experiences: Understanding Visions of the Afterlife.* Oxford UP, 2016. Hall, Jacqueline Dowd, et al. *Like a Family: The Making of a Southern Cotton Mill World.* U of North Carolina P, 1987. ▶ 如果作者数量在三位或以上，在第一个作者后面使用"et al."（仿照上例 *Like a Family*）
	文内夹注	(Young and Zabriskie, *Black Theater* 15–26)或 (Young and Zabriskie 15–26) (Fischer and Mitchell-Yellin 15–18) (Hall et al. 67)
图书，多个版本	参考文献	Strunk, William, Jr., and E. B. White. *The Elements of Style.* 50th anniversary ed., Longman, 2009. Saller, Carol Fisher. *The Subversive Copy Editor: Advice from Chicago.* 2nd ed., U of Chicago P, 2016. Head, Dominic, editor. *The Cambridge Guide to Literature in English.* 3rd ed., Cambridge UP, 2006. ▶ 假设上例是一部多卷本著作，具体的卷数要写在版次后面（即，3rd ed., vol. 2）
	文内夹注	(Strunk and White 12) (Saller 97) (Head 15)
主编图书	参考文献	Ross, Jeffrey Ian, editor. *Routledge Handbook of Graffiti and Street Art.* Routledge, 2016. Gilbert, Sandra, and Susan Gubar, editors. *Feminist Literary Theory and Criticism: A Norton Reader.* Norton, 2007. Cheng, Jim, et al., editors. *An Annotated Bibliography for Taiwan Film Studies.* Columbia UP, 2016. ▶ 如果有三位或以上编者，只写一位，然后在后面使用"et al."
	文内夹注	(Ross 93) (Gilbert and Gubar 72) (Cheng et al. 12)

续表

图书，匿名作者或无作者	参考文献	*Through Our Enemies' Eyes: Osama Bin Laden, Radical Islam, and the Future of America.* Brassey's, 2003. *The Secret Lives of Teachers.* U of Chicago P, 2015. ▶ 不要为匿名作者标注 Anonymous。由于缺少作者信息，排序时以标题的首字母为依据（但需要忽略标题开头的 A/An/The），所以 *The Holy Koran* 按字母顺序排在字母 H 之下
	文内夹注	(*Through Our Enemies' Eyes*) (*Secret Lives*)
电子图书	参考文献	Miranda, Lin-Manuel, and Jeremy McCarter. *Hamilton: The Revolution.* Kindle ed., Grand Central Publishing, 2016. Auspitz, Kate. *Wallis's War: A Novel of Diplomacy and Intrigue.* iBooks ed., U of Chicago P, 2015. ▶ MLA 格式将电子书视为图书的一个版本，因此标注了"ed."一词（edition 的缩写）
	文内夹注	(Miranda and McCarter) (Auspitz, ch. 3) ▶ 如果你引用的电子书无法确定准确的页码或属于自适应的排版格式，则需提供章节或其他标志信息
在线图书	参考文献	James, Henry. The Turn of the Screw. Martin Secker, 1915. *Internet Archive*, archive.org/details/in.ernet.dli.2015.95031. Skrentny, John David. *After Civil Rights: Racial Realism in the New American Workplace.* Princeton UP, 2014. *eBook Academic Collection, EBSCOhost.* ▶ 需要在图书的出版信息后面附上它所在的网站和网址 ▶ 如果图书来自图书馆的数据库，文献的获取链接会非常长，此时你可以选择不写网址
	文内夹注	(James 10) (Skrentny 105)
多卷本著作	参考文献	Cunningham, Noble E., Jr., editor. *Circular Letters of Congressmen to Their Constituents, 1789–1829.* Omohundro Institute of Early American History and Culture / U of North Carolina P, 2013. 3 vols. ▶ 若存在不止一家出版商，可以都列入文献，用斜线隔开
	文内夹注	(Cunningham)或(Cunningham 3: 21) ▶ 指的是第三卷第 21 页 (Cunningham, vol. 3) ▶ 如果选择引用整卷书，要用"vol."标注，避免读者误将数字认作页码
多卷本中的单卷作品	参考文献	Caro, Robert A. *The Passage of Power.* 2012. *The Years of Lyndon Johnson*, vol. 4, Knopf, 1982–. 5 vols. projected. Iriye, Akira. *The Globalizing of America.* 1993. *Cambridge History of American Foreign Relations*, edited by Warren I. Cohen, vol. 3, Cambridge UP, 1993. 4 vols. ▶ 鉴于单卷作品都有各自的标题，而且 MLA 格式不要求注明未使用的同系列作品，所以你也可以只标注作品的标题，省去卷号
	文内夹注	Caro, Robert A. *The Passage of Power.* Knopf, 2012. Iriye, Akira. *The Globalizing of America.* Cambridge UP, 1993. (Caro) (Iriye)

续表

早期版本的重印版	参考文献	Barzun, Jacques. *Simple and Direct: A Rhetoric for Writers*. 1985. U of Chicago P, 1994. Smith, Adam. *An Inquiry into the Nature and Causes of the Wealth of Nations*. 1776. Edited by Edwin Cannan, U of Chicago P, 1976.
	文内夹注	(Barzun, *Simple*)或(Barzun) (Smith, *Inquiry*)或(Smith)
译著	参考文献	Gan, Aleksei. *Constructivism*. 1922. Translated by Christina Lodder, Editorial Tenov, 2014. Tocqueville, Alexis de. *Democracy in America*. Edited by J. P. Mayer, translated by George Lawrence, Harper, 2000. ▶ 译者和原作者按其在图书扉页上出现的顺序排列 *Beowulf: A New Verse Translation*. Translated by Seamus Heaney, Farrar, 2000. ▶ *Beowulf*(《贝奥武夫》)的作者不详,译者写在了书名后面。但也有例外,如果你希望对某个特定译本发表评述,可以把译者名字放在最前面,比如: Heaney, Seamus, translator. *Beowulf: A New Verse Translation*. Farrar, 2000. Lodder, Christina, translator. *Constructivism*. By Aleksei Gan, 1922. Editorial Tenov, 2014.
	文内夹注	(Gan, *Constructivism*)或(Gan) (Tocqueville, *Democracy in America*)或(Tocqueville) (Heaney, *Beowulf*)或(*Beowulf*) (Lodder)
主编图书中的一章	参考文献	Jones, Kima. "Homegoing, AD." *The Fire This Time: A New Generation Speaks about Race*, edited by Jesmyn Ward, Scribner, 2016, pp. 15–18.
	文内夹注	(Jones 16)
期刊论文	参考文献	Sampsell, Gary. "Popular Music in the Time of J. S. Bach: The Leipzig Mandora Manuscript." *Bach*, vol. 48, no. 1, 2017, pp. 1–35. Domínguez Torres, Mónica. "Havana's Fortunes: 'Entangled Histories' in Copley's *Watson and the Shark*." *American Art*, vol. 30, no. 2, Summer 2016, pp. 8–13. ▶ 如果论文原标题内部带有双引号,引用时改为单引号 Ünal, Yusuf. "Sayyid Quṭb in Iran: Translating the Islamic Ideologue in the Islamic Republic." *Journal of Islamic and Muslim Studies*, vol. 1, no. 2, Nov. 2016, pp. 35–60, https://doi.org/10.2979/jims.1.2.04. Judd, David, and James Rocha. "Autonomous Pigs." *Ethics and the Environment*, vol. 22, no. 1, Spring 2017, pp. 1–18, https://doi.org/10.2979/ethicsenviro.22.1.01. ▶ 相比于普通网址,更推荐使用 DOI 编码。MLA 格式中的网址不需要在开头标注"https://",但是对于 DOI 编码,建议在前面添加"https://doi.org/" 如果作者人数达到三位或以上,可以写作:Judd, David, et al. Tyler, Tom. "Snakes, Skins and the Sphinx: Nietzsche's Ecdysis." *Journal of Visual Culture*, vol. 5, no. 3, Dec. 2006, pp. 365–85. Abstract. ▶ 如果只涉及引用文章的摘要,只需在文章信息后面加上"Abstract"一词
	文内夹注	(Sampsell)或(Sampsell 31) (Domínguez Torres 12) (Ünal)或(Ünal 58)或(Ünal, "Sayyid Quṭb" 58) ▶ 若涉及同一作者的不同作品,需标记论文的标题以示区分 (Judd and Rocha 3–5) (Tyler)

期刊论文，非英语	参考文献	Joosten, Jan. "Le milieu producteur du Pentateuque grec." *Revue des Études Juives*, vol. 165, nos. 3–4, juillet- décembre 2006, pp. 349–61. ▶ 如果页码范围的前后两个数字都在 99 以内，则应完整写出。其他情况下，只要不影响理解，允许后面的数字只写最后两位
	文内夹注	(Joosten)或(Joosten 356)
报纸或杂志的文章	参考文献	"Retired U.S. General Is Focus of Inquiry over Iran Leak." *The New York Times*, 28 June 2013, p. A18. ▶ 如果不清楚作者是谁，允许以标题开头 Halbfinger, David M. "Politicians Are Doing Hollywood Star Turns." *The New York Times*, 6 Feb. 2007, natl. ed., pp. B1+. ▶ 采用欧式日期标注，即"日月年" ▶ 对于一篇连续占据 B1、B2、B3 三个页面的文章来说，页码范围应写成 B1–B3。而上例中的加号说明：该文章的页码不是连续的，剩余部分刊登在其他版面（实际刊登在 B7 版） Halper, Evan. "Congress Takes Aim at the Clean Air Act, Putting the Limits of California's Power to the Test." *Los Angeles Times*, 3 Aug. 2017, 3:00 a.m. PDT, www.latimes.com/ politics/la-na-pol-smog-republicans-20170803-story.html. ▶ 如果文章频繁更新，请附上文章的发布时间（时间戳） ▶ 网络来源的文章无需注明页码
	文内夹注	("Retired U.S. General" A18) (Halbfinger B7)或为了区分同时引用的 Halbfinger 的其他文章，将夹注写成 (Halbfinger, "Politicians" B7) (Halper)
书评	参考文献	Pugh, Allison J. Review of *The Gender Trap: Parents and the Pitfalls of Raising Boys and Girls*, by Emily W. Kane. *American Journal of Sociology*, vol. 119, no. 6, May 2014, pp. 1773–75. Paxton, Robert O. Review of *The Nazi-Fascist New Order for European Culture*, by Benjamin G. Martin. *New York Review of Books*, 26 Oct. 2017, www.nybooks.com/articles/2017/10/26/nazi-fascist-cultural-axis.
	文内夹注	(Pugh 1774)或(Pugh, "Gender Trap" 1774) (Paxton)或(Paxton, "Nazi-Fascist New Order")
未发表的论文、毕业论文或学位论文	参考文献	Levine, Daniel H. "What Pope Francis Brings to Latin America." CLALS Working Paper Series 11, 12 Apr. 2016, papers.ssrn.com/sol3/papers.cfm?abstract_id=2761467. Sierra, Luis Manuel. *Indigenous Neighborhood Residents in the Urbanization of La Paz, Bolivia, 1910–1950*. 2013. State U of New York at Binghamton, PhD dissertation, ProQuest, search.proquest.com/docview/1507870239?accountid=14657.
	文内夹注	(Levine) (Sierra)或(Sierra 49)
档案材料、手稿集	参考文献	Franklin, Isaac. Letter to R. C. Ballard. 28 Feb. 1831, Rice Ballard Papers. Southern Historical Collection, Wilson Lib. U of North Carolina, Chapel Hill. MS. Series 1.1, folder 1. Lamson, Mary Swift. "An Account of the Beginning of the B.Y.W.C.A." 1891, Manuscript. Boston YWCA Papers. Schlesinger Lib, Radcliffe Institute for Advanced Study, Harvard U. Cambridge, MA.

档案材料、手稿集	参考文献	Szold, Henrietta. Letter to Rose Jacobs. 3 Feb. 1932. Rose Jacobs-Alice L. Seligsberg Collection. Judaica Microforms. Brandeis Lib. Waltham, MA. Microform, reel 1, book 1. Szold, Henrietta. Letter to Rose Jacobs. 9 Mar. 1936. Central Zionist Archives, Jerusalem. A/125/112. Copland, Aaron. "At the Thought of Mozart." Manuscript / mixed material, Library of Congress, Washington, DC, www.loc.gov/item/copland.writ0036. Accessed 13 Oct. 2017.
	文内夹注	(Franklin)或(Franklin to Ballard)或(Franklin to Ballard, 28 Feb. 1831) (Lamson)或(Lamson 2) (Szold)或(Szold to Jacobs)或(Szold to Jacobs, 3 Feb. 1932) (Szold)或(Szold to Jacobs)或(Szold to Jacobs, 9 Mar. 1936) (Taraval)或(Taraval, par. 23) ▶ 上例引用的手稿使用的不是页码，而是段落编号 (Taft)或(Taft 149)
百科全书、词典	参考文献	*Encyclopaedia Britannica*, 15th ed., 1987. *Merriam-Webster*, www.merriamwebster.com. *Oxford English Dictionary*, 2nd ed., 1989. ▶ 这里引用的是整部作品。具体涉及的条目要在正文中提及 ▶ 需要为纸质版的百科全书或词典标注版次和年份 ▶ 如果这些作品是由众所周知的出版商发行的，引用时可以省略出版商信息 如果你希望在此处列出参考的具体条目，可以写作： "App." *Merriam-Webster*. www.merriam-webster.com/dictionary/app. Accessed 24 Oct. 2017 ▶ 如果不清楚条目的创建时间，需要列出访问日期 "pluck, n.1." *OED Online*, www.oed.com/view/Entry/145996. Accessed 24 Oct. 2017. ▶ 名词"pluck"有两个义项，这里引用的是第一个含义，所以标注了"n.1."。pluck 另一个名词含义指的是一种鲜为人知的鱼 Kania, Andrea. "The Philosophy of Music." Revised 11 June 2017. Stanford *Encyclopedia of Philosophy*, plato.stanford.edu/entries/music.
	文内夹注	("App")或(*Merriam-Webster*) ("Pluck")或(*Oxford English Dictionary*) (Kania)
《圣经》《古兰经》	参考文献	*Tanakh: The Holy Scriptures: The New JPS Translation according to the Traditional Hebrew Text.* Jewish Publication Society, 1985. ▶ 对《圣经》《古兰经》等宗教典籍的引用通常不包括在参考文献中，除非你想表明你使用了某个特定版本或译本
	文内夹注	(*Tanakh*, Genesis 1.1, 1.3–5, 2.4) 如果后文继续引用这个版本，可以直接写(Genesis 1.1, 1.3–5, 2.4) ▶ 圣经的每卷书可以缩写（以创世记为例：Gen. 1.1, 1.3–5, 2.4) ▶ 接下来的四卷书缩写为 Exod.（出埃及记）、Lev.（利未记）、Num.（民数记）以及 Deut.（申命记）。如果你需要查询更多信息，在网络中搜索"abbreviations + Bible"即可 (Koran 18.65–82)

演讲、学术报告或课程讲座	参考文献	Zien, Katherine. "Good Neighbor, Good Soldier: Staging Transhemispheric Militarization in the Former Panama Canal Zone." Paper presented at the ASTR/TLA Annual Conference, 4 Nov. 2016, Minneapolis Marriott City Center. Skeel, David. "Is Justice Possible?" Lecture, 13 Nov. 2014, U of Hong Kong. *YouTube*, uploaded by Faith HKU, 3 Dec. 2014, youtu.be/qeR_Y2KQTts. Doniger, Wendy. Course on evil in Hindu mythology. 15 Mar. 2007, U of Chicago. Lecture. ▶ 如果无法直观地看出讲话的形式（live lecture/address），可在完整的条目信息后面标注
	文内夹注	(Zien) (Skeel) (Doniger)
私人通信或访谈	参考文献	▶ 私下的沟通一般只在文中提及就可以了。如果希望写入参考文献，可仿照如下格式： Alter, Robert. Personal interview. 21 Oct. 2015. Sarkozy, Nicolas. Telephone interview. 5 May 2017. Lucas, George. Video interview, Skype. 18 Feb. 2010. Anonymous US soldier, recently returned from Afghanistan. Personal interview. 28 Jan. 2014. Coates, Ta-Nehisi. "On Charlottesville, Trump, the Confederacy, Reparations & More." Interview by Amy Goodman and Juan González. *Democracy Now!*, 15 Aug. 2017, www.democracynow.org/2017/8/15/full_interview_ta_nehisi_coates_on. Rosenquist, James. "Reminiscing on the Gulf of Mexico: A Conversation with James Rosenquist." Interview by Jan van der Marck. *American Art*, vol. 20, no. 3, Fall 2006, pp. 84–107.
	文内夹注	(Alter) (Sarkozy) (Lucas) (anonymous US soldier) (Coates) (Rosenquist 93)
诗歌	参考文献	Auden, W. H. "The Shield of Achilles." 1952. *Collected Poems*, edited by Edward Mendelson, Random House, 2007, pp. 596–98. ▶ 1952 是 Mendelson 为 Auden 这首诗标注的年份。在诗歌全集或选集中，允许不标注原始的发表时间 Robards, Brooks. "The Front." *On Island: Poems and Paintings of Martha's Vineyard*, Summerset Press, 2014, p. 17. Balakian, Peter. "Near the Border." *Ozone Journal*, U of Chicago P, 2015, pp. 69–74. Paredez, Deborah. "St. Joske's." 2012. *Poets.org*, www.poets.org/poetsorg/poem/st-joskes.
	文内夹注	(Auden 596)或(Auden, "Shield of Achilles" 596)或("Shield of Achilles" 596)或(Auden 596; lines 9–11)或(Auden, lines 9–11)或(lines 9–11)或(9–11) ▶ 不要把 line 或 lines 用缩写取代，因为小写字母 l 容易同数字 1 混淆 (Robards) (Balakian 70)或(Balakian, "Near the Border" 70) (Paredez)或(Paredez, "St. Joske's")

续表

戏剧剧本	参考文献	Ruhl, Sarah. *Stage Kiss*. 2011. Theatre Communications Group, 2014. ▶ 该剧于 2011 年首次上演，于 2014 年出版发行 Shakespeare, William. *Romeo and Juliet*. ▶ 若想引用某个具体版本： Shakespeare, William. *Romeo and Juliet*. Edited by Brian Gibbons, Methuen, 1980. ▶ 如果是在线版本： Shakespeare, William. *Romeo and Juliet*. Edited by William J. Rolfe, American Book, 1907. Project Gutenberg, www.gutenberg.org/files/47960/47960-h/47960-h.htm.
	文内夹注	(Ruhl)或(Ruhl, *Stage Kiss*) (Shakespeare, *Romeo and Juliet* 1.3.12–15)或(*Romeo and Juliet* 1.3.12–15)或行文中已经提到剧名的，可以直接写(1.3.12–15) ▶ 此处指的是第一幕第三场，第 12 至 15 行。各项信息之间用英文句点隔开 ▶ 如果在文中多次引用莎士比亚的某部戏剧，可根据 MLA 格式的规定对剧目进行缩写（*Hamlet* 的缩写是 *Ham.*）。可以在第一次提到 *Romeo and Juliet* 时说明其缩写形式为 "*Rom.*"，接下来的文内夹注就都用缩写，例如(*Rom.* 1.3.12–15)
戏剧或舞蹈演出	参考文献	*Romeo and Juliet*. Choreography by Krzysztof Pastor, music by Sergei Prokofiev, Joffrey Ballet, 13 Oct. 2016, Auditorium Theatre, Chicago. *The Rembrandt*. Written by Jessica Dickey, directed by Hallie Gordon, performance by Francis Guinan and John Mahoney, 6 Oct. 2017, Steppenwolf Theatre, Chicago. ▶ 戏剧、音乐、舞蹈等合作式的演出有扮演各种角色的工作人员，如果你的文章格外关注某个人物或某个职位，就把那个人写在最前面。例如，希望强调 Francis Guinan 的表演： Guinan, Francis, performer. *The Rembrandt*. Written by Jessica Dickey, directed by Hallie Gordon, 6 Oct. 2017, Steppenwolf Theatre, Chicago.
	文内夹注	(*Romeo and Juliet*) (*Rembrandt*)或(Guinan)
电视节目	参考文献	"The Children." *Game of Thrones*, written by David Benioff and D. B. Weiss, directed by Alex Graves, season 4, episode 10, HBO, 15 June 2014. "Pressing the Flesh." *Scandal*, written by Shonda Rhimes, performance by Kerry Washington, Katie Lowes, and Scott Foley, season 7, episode 2, ABC, 12 Oct. 2017. Hulu, www.hulu.com/watch/1156610.
	文内夹注	("The Children") ("Pressing the Flesh")
电影	参考文献	*Godfather II*. Directed by Francis Ford Coppola, performance by Al Pacino, Robert De Niro, Robert Duvall, and Diane Keaton, screenplay by Francis Ford Coppola and Mario Puzo, based on the novel by Mario Puzo, Paramount Pictures, 1974. Paramount Home Video, Godfather DVD Collection, 2003. ▶ 可以灵活处理的是演员、制片人、编剧、剪辑、摄像等人物信息，即根据实际写作需求决定是否列出，并按照相对于论文的重要性进行排序，把涉及的姓名写在片名和发行商之间 ▶ 如需着重强调某个人物的工作，可将姓名和职责（比如 performer）写到片名之前：

续表

电影	参考文献	Coppola, Francis Ford, director. *Godfather II*. Performance by Al Pacino, Robert De Niro, Robert Duvall, and Diane Keaton, Paramount Pictures, 1974. Paramount Home Video, Godfather DVD Collection, 2003. *Tig*. Directed by Kristina Goolsby and Ashley York, Beachside / Netflix, 2015. Netflix, www.netflix.com/title/80028208. ▶ 若存在不止一家制作单位，可以都列入文献，用斜线隔开。上例中，Netflix 既是制片方，也是影片的获取平台
	文内夹注	(*Godfather II*)或(Coppola) (*Tig*)
艺术作品	参考文献	Tintoretto, Jacopo Robusti. *The Birth of John the Baptist*. 1550s, State Hermitage Museum, St. Petersburg. Oil on canvas, 181 × 266 cm. ▶ 尺寸和艺术种类可以不写 Tintoretto, Jacopo Robusti. *The Birth of John the Baptist*. 1550s, State Hermitage Museum, St. Petersburg. *Tintoretto: Tradition and Identity*, by Tom Nichols, Reaktion Books, 1999, p. 47. ▶ 如果是复制品，还应标明它的出处 Tintoretto, Jacopo Robusti. *The Birth of John the Baptist*. 1550s, State Hermitage Museum, St. Petersburg, www.hermitagemuseum.org/wps/portal/hermitage/digital-collection/01.+Paintings/32113/?lng=en. ▶ 如果是在线的艺术作品，应写清网址或网站。如果网站与博物馆同名，则没必要重复写两次
	文内夹注	(Tintoretto)或(Tintoretto, *Birth of John the Baptist*)
照片	参考文献	Adams, Ansel. *Monolith, the Face of Half Dome, Yosemite National Park*. 1927, Art Institute, Chicago. Adams, Ansel. *Dunes, Oceano, California*. 1963. *MoMA*, www.moma.org/collection/works/58283.
	文内夹注	(Adams)或(Adams, *Monolith*) (Adams)或(Adams, *Dunes, Oceano*)
图：地图、图表、曲线图或表格	参考文献	Digout, Delphine. "Climate Change Vulnerability in Africa." Revised by Hugo Ahlenius, UNEP/GRID-Arendal, 2002. *GRID-Arendal*, old.grida.no/graphicslib/detail/climate-change-vulnerability-in-africa_7239. "Presidential Election Results: Donald J. Trump Wins." 9 Aug. 2017. *The New York Times*, www.nytimes.com/elections/results/president. Interactive map. ▶ 如果不能直观地看出图形的格式（上例中的 interactive map），可进一步完善标注 "1427 E. 60th St., Chicago IL 60637." *Google Maps*, www.google.com/maps/place/The+University+of+Chicago+Press,+1427+E+60th+St,Chicago,+IL+60637/. Accessed 25 Oct. 2017. 2017. ▶ 如果线上的地图没有绘制日期，则应注明访问日期 ▶ 从图书、期刊中引用的地图通常要在文章的正文中加以说明（请看下面"文内夹注"的部分），而参考文献里列出的应该是整个作品 Menkhaus, Ken. "Governance without Government in Somalia: Spoilers, State Building, and the Politics of Coping." *International Security*, vol. 31, no. 7, Winter 2006/7, pp. 74–106.
	文内夹注	("Climate Change Vulnerability in Africa") ("Presidential Election Results") ("Map of Somalia, 2006/7")或(Menkhaus 79, map)

续表

音乐唱片	参考文献	Johnson, Robert. "Come On in My Kitchen (Take 1)." 1936. *Robert Johnson: King of the Delta Blues Singers*, expanded ed., Columbia/Legacy, 1998. Johnson, Robert. "Traveling Riverside Blues." 1937. *Robert Johnson: King of the Delta Blues Singers*, Columbia Records 1654, 1961. LP. Allman Brothers Band. "Come On in My Kitchen." Written by Robert Johnson, *Shades of Two Worlds*. Sony, 1991. Barber, Samuel. Cello sonata, for cello and piano, op. 6. *Barber: Adagio for Strings, Violin Concerto, Orchestral and Chamber Works*, St. Louis Symphony, conducted by Leonard Slatkin, Alan Stepansky, cello, Israela Margalit, piano, EMI Classics 74287, 2001. CD. ▶ 曲目编号也可一并列入发行信息里，方便读者找到特定的乐曲 ▶ 如果引自实体唱片，应在条目的最后标明 ▶ 如需强调某个人的角色（例如 pianist），就把他/她写在最前面： Margalit, Israela, piano. Cello sonata, for cello and piano, op. 6. *Barber: Adagio for Strings, Violin Concerto, Orchestral and Chamber Works*, St. Louis Symphony, conducted by Leonard Slatkin, Alan Stepansky, cello, EMI Classics 74287, 2001. Beyoncé. "Formation." *Lemonade*, Parkwood Entertainment, 2016. ▶ 无需注明你听音乐的软件或网站。如果你希望列出，参考文献的格式为： Beyoncé. "Formation." *Lemonade*, iTunes app, Parkwood Entertainment, 2016.
	文内夹注	(Johnson)或(Johnson, "Come") (Johnson)或(Johnson, "Traveling") (Allman Brothers)或(Allman Brothers, "Come") (Barber)或(Barber, Cello sonata) (Margalit)或(Margalit, Cello sonata) (Beyoncé)或(Beyoncé, "Formation")
音乐录像带	参考文献	Arcade Fire. *Everything Now*. YouTube, uploaded by ArcadeFireVEVO, 1 June 2017, youtu.be/zC30BYR3CUk. Rihanna et al. *FourFiveSeconds*. Directed by Inez & Vinoodh, produced by Stephanie Bargas and Jeff Lepine. Vevo, 3 Feb. 2015, www.vevo.com/watch/rihanna/fourfiveseconds/QMFUA1590146. Video. ▶ 若从条目本身看不出文件类型，可以增加"Video"一词 ▶ 参考文献中不引用观众发表的评论（或任何网页内容）
	文内夹注	(Arcade Fire)或(Arcade Fire, *Everything Now*) (Rihanna et al.)
活页乐谱	参考文献	Bach, Johann Sebastian. *Toccata and Fugue in D Minor*. 1708. BWV 565, arranged by Ferruccio Benvenuto Busoni for solo piano, G. Schirmer LB1629, 1942. ▶ 这首乐曲写于 1708 年，具有 Bach 作品的标准分类编码 BWV 565。这个改编是由 G. Schirmer 在 1942 年发行的，Schirmer 对它的编号是 LB1629
	文内夹注	(*Toccata and Fugue in D Minor*)或(Bach, *Toccata and Fugue in D Minor*)
唱片内页文字说明	参考文献	Reich, Steven. Liner notes. *Different Trains*. Kronos Quartet, Elektra/Nonesuch 9 79176-2, 1988. CD.
	文内夹注	(Reich, *Different Trains*)
有声书	参考文献	Eliot, George. *Middlemarch*. Narrated by Juliet Stevenson, Audible, 2011. Audiobook.

有声书	文内夹注	(Eliot)或(Eliot, *Middlemarch*)或(Eliot 01:02:20–27) ▶ 若想引用特定的文字片段，可以通过文内夹注指出具体的时间点或时间段
广告	参考文献	Advertisement for *Letters from Iwo Jima. The New York Times*, 6 Feb. 2007, p. B4. Advertisement for Vitamin Water. *Rolling Stone*, 15 June 2017, p. 17. "Be a Mindsticker," advertisement for Tab cola. Coca-Cola Co., circa late 1960s. *Dailymotion*, 8 Aug. 2007, www.dailymotion.com/video/x2s3qd.
	文内夹注	(*Letters from Iwo Jima* advertisement) (Vitamin Water advertisement) (Tab advertisement)
政府文献	参考文献	United States, Congress, Senate, Committee on Armed Services. *Hearings on S. 758, a Bill to Promote the National Security by Providing for a National Defense Establishment*. Government Printing Office, 1947. 80th Congress, 1st session, Senate Report 239. United States, Congress, House. *American Health Care Act. Congress.gov*, 2017, www.congress.gov/bill/115th-congress/house-bill/1628/text. ▶ 未必一定要列出美国国会的届数、会议次数（第一/第二次会议）、议院名称（众/参议院）、文件的编号或类型等信息 United States, Environmental Protection Agency, Office of Air and Radiation. *A Brief Guide to Mold, Moisture, and Your Home*. EPA. gov, 2012, www.epa.gov/mold/brief-guide-mold-moisture-and-your-home. Freedman, Stephen. *Four-Year Impacts of Ten Programs on Employment Stability and Earnings Growth: The National Evaluation of Welfare-to-Work Strategies*. US Department of Education, 2000, www.mdrc.org/publication/four-year-impacts-ten-programs-employment-stability-and-earnings-growth. United States, Department of State. Daily Press Briefing. US Department of State, 9 Aug. 2017, www.state.gov/r/pa/prs/dpb/2017/08/273288.htm.
	文内夹注	(US Cong., Senate, Committee on Armed Services) ▶ 如果只参考了委员会的一项文件，文内引用时就无需标注听证会或报告的编号 (US Cong., Senate, Committee on Armed Services, *Hearings on S. 758*, 1947) ▶ 如果文中提到了委员会的多项文件，文内夹注就需要指明此处参考的是哪一份文件。但再次引用时，可以缩写： (*Hearings on S.* 758) (US, EPA, Office of Air and Radiation 15) (Freedman 6)或(Freedman, *Four-Year Impacts*) (US Dept. of State)或(US Dept. of State, *Press Briefing*, 9 Aug. 2017)
网站、网页	参考文献	"News and Advocacy." *American Historical Association*, www.historians.org/news-and-advocacy. Accessed 9 Nov. 2017. de Blasio, Bill. "State of the City 2017." *Office of the Mayor*, City of New York, 13 Feb. 2017, www1.nyc.gov/office-of-the-mayor/state-of-our-city.page. Lipson, Charles. "News and Commentary: US and the World." *Charles Lipson*, www.charleslipson.com/News-links.htm. Accessed 29 Aug. 2017. ▶ 如需引用没有明确作者的网页，可以把网页或网站的标题写在最前面 ▶ 如果无法注明出版、发布日期或修订日期，则需注明访问日期 ▶ 网页或在线文档可能无法列出页码，但你可以尝试标明小节（Lipson, sec. 7）或段落（Lipson, pars. 3–5）

续表

网站、网页	文内夹注	("News and Advocacy") (de Blasio)或(de Blasio,"State of the City") (Lipson)
博客文章或评论	参考文献	Jayson, Sharon."Is Selfie Culture Making Our Kids Selfish?" *Well, The New York Times*, 23 June 2016, well.blogs.nytimes.com/2016/06/23/is-selfie-culture-making-our-kids-selfish. Arana, Ana."The Deep Magic of Mexico." *The Big Roundtable*, 8 Dec. 2016, thebigroundtable.com/the-deep-magic-of-mexico-2bf1b341d1a1. ▶ 博客文章或发帖的引用方式可以仿照报纸或新闻网站文章的引用方式执行 Cheryl. Comment on "Is Selfie Culture Making Our Kids Selfish?" *Well, The New York Times*, 23 June 2016, well.blogs.nytimes.com/2016/06/23/is-selfie-culture-making-our-kids-selfish.
	文内夹注	(Jayson)或(Jayson, "Selfie Culture") (Arana)或(Arana, "Deep Magic")
视频	参考文献	*Duck and Cover*. Archer Productions / Federal Civil Defense Administration, 1951. *YouTube*, uploaded by Nuclear Vault, 11 July 2009, youtu.be/IKqXu-5jw60. ▶ 这部1951年的电影于2009年在网站上线 "Jury Convicts Martin Shkreli on Three Counts of Security Fraud." NBC, 4 Aug. 2017, www.nbcnews.com/video/jury-convicts-martin-shkreli-on-three-counts-of-securities-fraud-1017719875664. Wang, Tricia. "The Human Insights Missing from Big Data." TED, Sept. 2016, www.ted.com/talks/tricia_wang_the_human_insights_missing_from_big_data. Video. ▶ 若从条目本身看不出文件类型，可以增加"Video"一词。 ▶ 为什么 *Duck and Cover* 用了斜体字，而 Jury Convicts Martin Shkreli 和 Human Insights 却放在双引号里呢？MLA 格式要求将电影或电视剧的名称用斜体字表示，单集标题使用双引号。就目前的情况而言，新媒体形式层出不穷，上面规则所划的界限也没那么明确了。*Duck and Cover* 虽然时长不足，但是是以独立电影的方式制作的，因此写成斜体。后面两个例子更像是一系列视频中的一集
	文内夹注	(*Duck and Cover*) ("Jury")
多媒体应用(电子游戏以及其他独立软件)	参考文献	*Madden NFL 18*. Standard ed., EA Sports, 25 Aug.2017. ▶ 可以包含其他你认为必要的信息： *Madden NFL 18*. EA Sports, PlayStation 4, Nov. 2017. *Geocaching*. Groundspeak, v. 5.6.1, 2017.
	文内夹注	(*Madden NFL 18*) (*Geocaching*)
播客	参考文献	"Has Anybody Called You?" *S-Town*, ch. 2, hosted by Brian Reed, produced by Brian Reed and Julie Snyder, Serial Productions, 28 Mar. 2017, stownpodcast.org/chapter/2. ▶ 或 Reed, Brian, host. "Has Anybody Called You?" *S-Town*, ch. 2, produced by Brian Reed and Julie Snyder, Serial Productions, 28 Mar. 2017, stownpodcast.org/chapter/2. DeGeneres, Ellen. "Ellen's Parenting Advice." *The Ellen Show Podcast*, Telepictures Productions, 26 June 2015, iTunes.

续表

播客	文内夹注	("Has Anybody Called You?")或(Reed) (DeGeneres)或(DeGeneres, "Ellen's Parenting Advice")
社交媒体 (Facebook, Instagram, Twitter)	参考文献	Zuckerberg, Mark. Profile. *Facebook*, www.facebook.com/zuck. Tyson, Neil deGrasse (@neiltyson). "Always seemed to me that millipedes have more legs than are necessary." *Twitter*, 7 July 2017, 1:25 p.m., twitter.com/ neiltyson/status/883421859072458752. ▶ 标题的位置可以使用社交媒体的帖文，最多截取160字符 Chicago Manual of Style. "Is the world ready for singular they? We thought so back in 1993." *Facebook*, 17 Apr. 2015, www.facebook.com/ChicagoManual/posts/10152906193679151. Richards, Corey (@coryrichards). "A quiet if not solemn day on Everest." *Instagram*, 24 May 2017, www.instagram.com/p/BUekBZLgDOW/. ▶ 以上形式适用于对公众可见的社交账号内容。如果引用的是私信等私密内容，可以仿照"短信"的引用方法
	文内夹注	(Zuckerberg) (Tyson) (Chicago Manual of Style) (Richards)或(Richards, "Quiet")
邮件、短信 （私信）	参考文献	▶ 短信或其他类型的私聊信息只需在正文中引注即可。如果一定要写入参考文献，可以使用下面的格式： Leis, Kathy. "Re: New Orleans family." Received by Karen Turkish, 3 Mar. 2017. ▶ 标题的位置可以填写邮件的主题 Lipson, Michael. *Facebook* message to Jonathan Lipson. 9 Feb. 2016. ▶ 根据论文的写作需要，可以考虑是否写清邮件或即时信息的时间（比如：9 Feb. 2010, 3:15 p.m.）
	文内夹注	(Leis) (Lipson)或(Lipson, M.)
电子论坛、邮件列表	参考文献	Polger, Mark Aaron. "CALL: Call for Proposals." American Library Association mailing list archives, ACRL Campaign for Libraries list, 10 May 2017, lists.ala.org/sympa/arc/academicpr/2017-05/msg00005.html. Ngo, Quinn. "Re: What Are Some Unknown Truths about College Life?" *Quora*, 14 July 2017, www.quora.com/What-are-some-unknown-truths-about-college-life.
	文内夹注	(Polger) (Ngo)

MLA 格式过去要求缩写的许多词汇，现在已经允许完整地写出来。还有一些术语依旧需要缩写。比如：editor 等词可以拼写完整，但 edition 必须缩写。除了 May/June/July，其他月份名称需要以缩写形式列出。表 7.3 提供了一些常用的缩写：

表7.3　MLA格式：常见的缩写

全称	缩写	全称	缩写	全称	缩写
and others	et al.	library	lib.	part	pt.
appendix	app.	note	n.	press	p
book	bk.	notes	nn.	pseudonym	pseud.
chapter	ch.	number	no.	专有名词中的 University	U
compare	cf.	opus	op.	University Press	UP
document	doc.	opuses	opp.	verse	v.
edition	ed.	page	p.	verses	vv.
especially	esp.	pages	pp.	versus	vs.
figure	fig.	paragraph	par.	volume	vol.

注：缩写都用小写字母（表示出版商名称的缩写词除外），后面通常紧跟英文句号。大部分缩写的复数形式为直接加 s，特例是 note（n.→nn.），opus（op.→opp.），page（p.→pp.）和 translator（复数为 trans.不变）。

引用诗歌时，不要把 line 或 lines 用缩写取代，因为小写字母 l 容易同数字 1 混淆。

出版社中常见 U、P、UP 等缩写，同时还可删掉 "Company" "Co." "Inc." 等表示 "公司" 的词。

7.2　MLA 格式的常见问题

（1）MLA 格式的 container 该如何理解？

container 就相当于承载出版信息（facts of publication）的容器。参考文献列表中的大多数条目都要列出作者、标题和某些出版信息（上述各项用英文句号切分，每个项目内部的分项用逗号隔开）。在出版信息的容器里，就要写上书目的出版商、出版日期，如果引用的是一集电视剧，需要写清剧目和其他相关信息。

在这些内容后面，有的时候要使用第二个 container——比如说你从网上引用了一本书或一期节目，在常规出版信息（第一个 container）后面，你可以补充网站名称和访问地址。两个 container 之间用句号隔开。

不是所有的网络资料都需要第二个 container。如果你从 *Los Angeles Times* 网站引用了一则新闻，而报纸跟网站的名称一致，都是《洛杉矶时报》，所以一个 container 就足够出示所有信息了。

上面的表格里有许多类似的例子。

（2）如果引文的原作者还引用了另一位作者，如何在文章里进行引注呢？

这种情况并不少见。以 Donald Kagan 所著的 *The Peloponnesian War* 为例，书中多次引用希腊历史学家 Thucydides 的作品。如果你使用 MLA 格式引用 Kagan，在行文时建议这么写：

Kagan approvingly quotes Thucydides, who says that Athens acquired this vital site "because of the hatred they already felt toward the Spartans" (quoted in Kagan 14).

针对这段文字，你需要在参考文献中列出 Kagan 的著作，不需要写出 Thucydides 的作品。但如果你论文的其他部分直接参考了 Thucydides 的著作，则应该如实体现在参考文献（Works Cited）中。

8

APA格式：社会科学、教育学和商学

APA 格式广泛用于心理学、教育学、商学以及社会科学（一些工科专业的学生也会用到 APA 格式，但他们更常使用第 13 章的 IEEE 格式或 ACSE 格式）。MLA 格式和 APA 格式都采用文内夹注，也都允许添加注释，并且注释需要用来辅助分析、评论，不能用于文献引注。二者的区别在于，APA 格式更侧重文献的出版时间，因此紧跟作者姓名的是年份信息。原因或许是，在 APA 格式适用的实证研究领域，人们看重知识和经验的不断累积，会更加关注成果的产生时间和先后顺序。

如需查阅 APA 格式的官方手册，请参考 2010 年出版的 *Publication Manual of the American Psychological Association*（第 6 版）和 2012 年出版的 *APA Style Guide to Electronic Resources*（第 6 版）。❶

首先，我们来看一组 APA 格式下的参考文献（APA 格式称之为 Reference List，MLA 格式对应的是 Works Cited，芝加哥引用格式对应 Bibliography）。下面三篇文献对应的分别是一篇期刊论文、主编图书中的一章和一部专著。

> Lipson, C. (1991). Why are some international agreements informal? *International Organization*, *45*, 495–538.
>
> Lipson, C. (1994). Is the future of collective security like the past? In G. Downs (Ed.), *Collective security beyond the cold war* (pp. 105–131). Ann Arbor: University of Michigan Press.
>
> Lipson, C. (2003). *Reliable partners: How democracies have made a separate peace*. Princeton, NJ: Princeton University Press.

观察可得，三篇文献都出自 C. Lipson 一人之手。针对这种情况，APA 格式要求将它们按照由远及近的时间顺序排列；如果某两个条目都是在 2018 年出版，则按著作标题的首字母排序，一条记为（2018a），另一条记为（2018b）。即将出版的著作可以在括号内注

---

❶ 2019 年 APA 格式发布了第 7 版。——译者注

明"in press"。还需要注意书名和文章标题的大写规则——APA格式把标题视为"句子"，所以只有"句首"的第一个字母大写；若标题含有冒号，冒号后第一个词的首字母大写。还有就是，专有名词的大写不受影响，这跟写句子也是一样的。

参考文献中涉及同一作者的独著在前，合著在后，即Pinker, S.所著的条目排在Pinker, S., & Jones, B.共同撰写的条目之前。在APA体系的文献列表中，多个作者之间用符号&连接，不要把"and"写出来。我也不清楚这么做的理由，就像足球比赛的时长规定一样，只需服从即可。

作者的姓氏写出来，名字都只保留首字母。这是APA格式有别于MLA或芝加哥格式的地方。另一个区别是，与二者相比，APA格式更频繁地使用逗号和括号。

如果通过论文正文无法辨认被引信息的来源，写夹注时，要附上作者信息，例如：

These claims are backed up by the most recent data (Nye, 2018).

如果行文时提到了原作者的名字，文内夹注可以省略作者。即：

Nye (2018) presents considerable data to back up his claims.

若出现了直接引用，文内夹注必须提供页码信息，比如：

"The policy is poorly conceived," according to Nye (2018, p. 12).

本章所列案例大多来自心理学、教育学、商学以及其他的社会科学领域，这些学科普遍采用APA格式，而我们已经讲过的MLA格式则聚焦在它所适用的人文学科领域。

我已经按照字母顺序将各类条目整理成表格（表8.1、表8.2），同时附上页码信息，方便你迅速定位相关内容。在本章的最后，我还列出了一些APA格式常用的缩写。

表8.1 APA格式索引表

文献类型	在本书中页码	文献类型	在本书中页码
摘要（abstract）	100	主编图书（edited）	98
广告（advertisement）	106	非英文著作（foreign language）	99
《圣经》（Bible）	103	多位作者（multiple authors）	97–98
博客（blog）	108	多个版本（multiple editions）	98
评论（comment）	108	多卷本著作（multivolume work）	99
文章、帖文（post）	108	无作者（no author）	98
图书（book）		单一作者（one author）	97
主编图书中的一章（chapter in edited book）	100	在线图书（online）	99
		早期版本的重印版（reprint）	99
机构作者（corporate author）	98	多本图书，同一作者（several by same author）	97
电子图书（e-book）	99		

续表

文献类型	在本书中页码	文献类型	在本书中页码
多卷本中的单卷作品（single volume in a multivolume work）	99	地图（map）	105
译著（translated volume）	99	多媒体应用（multimedia app）	109
图表（chart）	105	音乐唱片（musical recording）	106
古典著作（classical work）	104	报纸的文章（newspaper article）	
数据库或数据集（database or data set）	107	无作者（no author）	101
		有作者（with author）	101
诊断手册（diagnostic manual）	107–108	未发表的论文（paper, unpublished）	101–102
诊断性测试（diagnostic test）	107	私人通信（personal communication）	104
词典（dictionary）	103	照片（photograph）	105
私信（direct message）	110	播客（podcast）	109
学位论文或毕业论文（dissertation or thesis）	101–102	政策文件（policy paper）	103
电子论坛、邮件列表（electronic forum or mailing list）	110	墙报论文（poster session）	101–102
邮件（email）	110	预印本（preprint）	102
百科全书（encyclopedia）	103	《古兰经》（Qur'an）	103
脸书（Facebook）	110	书评（review）	101
图（figure）	105	社交媒体（social media）	110
电影（film）	104–105	软件（software）	109
政府文献（government document）	106	演讲、报告或讲座（speech, lecture, or talk）	104
曲线图（graph）	105	表格（table）	105
灰色文献（gray literature）	103	研究报告或技术报告（technical and research reports）	103
照片墙（Instagram）	110	电视节目（television program）	104
访谈（interview）	104	短信（text message）	110
期刊论文（journal article）		毕业论文（thesis）	101–102
非英语（foreign language）	101	推特（Twitter）	110
多位作者（multiple authors）	100		
单一作者（one author）	100	视频（video clip）	108–109
杂志的文章（magazine article）		网站、网页（website or page）	108
无作者（no author）	101		
有作者（with author）	101	白皮书（white paper）	103

表8.2 APA格式：参考文献和文内夹注

图书，单一作者	参考文献	King, B. J. (2017). *Personalities on the plate: The lives and minds of animals we eat*. Chicago, IL: University of Chicago Press. ▶ 即使是著名城市，APA 格式依然要求把它所在的州写出来❶。如果州名已经包含在出版社的名称中，就不必再重复列出（比如：Ann Arbor: University of Michigan Press） Naughton, B. (2007). *The Chinese economy: Transitions and growth*. Cambridge, MA: MIT Press. Fitzpatrick, M. (2017). *Invisible scars: Mental trauma and the Korean War*. Vancouver, British Columbia, Canada: University of British Columbia Press. ▶ 美国的州用了缩写，但加拿大的省名、地名是完整拼写出来的，同时还附上了国名 Canada
	文内夹注	(King, 2017) (Naughton, 2007) (Fitzpatrick, 2017)
多本图书、文章，同一作者	参考文献	Posner, R. C. (2007a). *Countering terrorism*. Lanham, MD: Rowman & Littlefield with the Hoover Institution. Posner, R. C. (2007b). *Economic analysis of law* (7th ed.). New York, NY: Aspen Law and Business. Posner, R. C. (2007c). *The little book of plagiarism*. New York, NY: Pantheon. Posner, R. C. (2006a). *Not a suicide pact: The Constitution in a time of national emergency*. New York, NY: Oxford University Press. Posner, R. C. (2006b). *Uncertain shield: The U.S. intelligence system in the throes of reform*. Lanham, MD: Rowman & Littlefield with the Hoover Institution. Posner R. C., & Becker, G. S. (2006). *Suicide and risk-taking: An economic approach*. Unpublished paper, University of Chicago. ▶ 注意：APA 格式不允许用破折号代替重复的作者名，每次都要把作者写出来 ▶ 如果作者相同、出版年份相同，文献条目按标题第一个重要实词的首字母排列，分别标记为 a、b、c。Posner 第三篇 2006 年的文章没有标记为 c，因为它的作者信息与前两条不同 ▶ Posner & Becker 合著的作品排在 Posner 独著的作品之后。若还引用了 Posner 跟其他作者合著的作品，按第二位作者姓氏的首字母排序即可
	文内夹注	(Posner, 2007a, 2007b, 2007c, 2006a, 2006b; Posner & Becker, 2006)❷
图书，多位作者	参考文献	Gazzaley, A., & Rosen, L.D. (2016). *The distracted mind: Ancient brains in a high-tech world*. Cambridge, MA: MIT Press. Useem, M., Singh, H., Neng, L., & Cappelli, P. (2017). *Fortune makers: The leaders creating China's great global companies*. New York, NY: Public Affairs. Butcher, T., Guo., X., Harris, J., Lette, K., Mankell, H., Moggach, D., ... Welsh, I. (2010). *Because I'm a girl*. New York, NY: Random House. ▶ 最多可列出七位作者的名字；如果是八位或以上，列出前六位和最后一位，中间使用省略号隔开❸

---

❶ 根据 APA 格式第 7 版手册，只列出出版商名称即可，地理位置不必再写入参考文献部分。——译者注
❷ 根据 APA 格式的要求，发表年份早的作品应该排在前面。——译者注
❸ 根据 APA 格式第 7 版手册，如果作者人数在二十人及以内，需要全部列出。若超过二十人，写出前十九位和最后一位作者姓名，二者之间用省略号隔开。——译者注

图书，多位作者	文内夹注	(Gazzaley & Rosen, 2016) (Useem, Singh, Neng, & Cappelli, 2017) 如果后文继续引用，可以直接写(Useem et al., 2017) ▶ 如果只有两位作者，需要全部列出。如果作者人数是三至五人，首次引用要全部列出，后续引用可以只写第一位作者，然后加上"et al."（Fubini et al., 2007）❶ ▶ 如果作者人数为六位或以上，所有夹注都采用第一位作者加"et al."的形式 ▶ 在括号内用&符号，但是在论文正文里的叙述不要用
图书，多个版本	参考文献	DiClemente, C. C. (2018). *Addiction and change: How addictions develop and addicted people recover* (2nd ed.). New York, NY: Guilford Press. ▶ 如果是修订版（revised edition），就删去版次"2nd ed."，在相同的位置写上"Rev. ed." Strunk, W., Jr., & White, E. B. (2009). The elements of style (50th anniversary ed.). New York, NY: Longman.
	文内夹注	(Strunk & White, 2009) ▶ 如果是直接引用他人原话，需要标注页码，格式为： (Strunk & White, 2009, p. 12)
图书，机构作者或无作者	参考文献	University of Michigan, Office of Student Publications. (2017). *2017 alumni directory*. Bloomington, IN: University Publishing Corporation. American Psychological Association. (2010). *Publication manual of the American Psychological Association* (6th ed.). Washington, DC: Author. ▶ 如果出版商跟机构作者相同，则在出版商的位置标注"Author" *The bluebook: A uniform system of citation* (20th ed.). (2015). Cambridge, MA: Harvard Law Review Association. ▶ 如果作品有多个版本，且没有作者，参考文献的格式为：*Title* (edition). (year). City, STATE: Publisher.
	文内夹注	(University of Michigan, 2017) ▶ 第一次引用要写完整。如果有明显机构简称，后续的文内夹注可以使用(American Psychological Association [APA], 2010) ▶ 后文再次引用，可以写为(APA, 2010)或(*Bluebook*, 2015)
主编图书	参考文献	Bakker, K. (Ed.). (2007). *Eau Canada: The future of Canada's water*. Vancouver, British Columbia, Canada: University of British Columbia Press. Bosworth, M., & Flavin, J. (Eds.). (2007). *Race, gender, and punishment: From colonialism to the war on terror*. New Brunswick, NJ: Rutgers University Press. Matthijs, M., & Blyth, M. (Eds.). (2015). *The future of the Euro*. Oxford, England: Oxford University Press.
	文内夹注	(Bakker, 2007) (Bosworth & Flavin, 2007) (Matthijs & Blyth, 2015)

---

❶ 根据 APA 格式第 7 版手册，正文引用三个及以上作者，全部用 et al.形式，无需在第一次引用时列出所有人。——译者注

续表

电子图书	参考文献	Toy, E. C., & Klamen, D. (2015). *Case files: Psychiatry* (5th ed.) [Kindle Fire HDX version]. Retrieved from http://www.amazon.com ▶ APA 格式建议将一些重要但是非例行的信息写在标题（卷数、版本）后面的方括号内❶ ▶ 电子检索信息（即网址）放在了出版商的位置。如果访问地址过长，可以只写网站主页（如上例所示）❷
	文内夹注	(Toy & Klamen, 2015)
在线图书	参考文献	Reed, J. (1922). *Ten days that shook the world* [Etext 3076]. Retrieved from http://www.gutenberg.org/dirs/etext02/10daz10.txt ▶ APA 格式不需要在网址后面添加英文句号，这与其他引用格式不同 ▶ APA 格式对访问日期不做要求，除非该资料频繁变化（例如 *Wikipedia* 上的文章）
	文内夹注	(Reed, 1922)
多卷本著作	参考文献	Rothschild, B. (2017). *The body remembers* (Vols. 1–2). New York, NY: Norton. Pflanze, O. (1963–1990). *Bismarck and the development of Germany* (Vols. 1–3). Princeton, NJ: Princeton University Press.
	文内夹注	(Rothschild, 2017) (Pflanze, 1963–1990)
多卷本中的单卷作品	参考文献	Rothschild, B. (2017). *The body remembers: Vol. 2. Revolutionizing trauma treatment.* New York, NY: Norton. Pflanze, O. (1990). *Bismarck and the development of Germany: Vol. 3. The period of fortification, 1880–1898.* Princeton, NJ: Princeton University Press.
	文内夹注	(Rothschild, 2017) (Pflanze, 1990)
早期版本的重印版	参考文献	Smith, A. (1976). *An inquiry into the nature and causes of the wealth of nations.* E. Cannan (Ed.). Chicago, IL: University of Chicago Press. (Original work published 1776) ▶ 最后的括号后面没有句号
	文内夹注	(Smith, 1776/1976)
译著	参考文献	Weber, M. (1958). *The Protestant ethic and the spirit of capitalism.* T. Parsons (Trans.). New York, NY: Charles Scribner's Sons. (Original work published 1904–1905)
	文内夹注	(Weber, 1904–1905/1958)
非英文著作	参考文献	Weber, M. (2005). *Die protestantische Ethik und der Geist des Kapitalismus* [The Protestant ethic and the spirit of capitalism]. Erftstadt, Germany: Area Verlag. (Original work published 1904–1905)
	文内夹注	(Weber, 1904–1905/2005)

---

❶ 根据 APA 格式第 7 版手册，格式、平台或者设备无需特意标注。如果你引用的有声书（Audiobook）是原文的简略本，或你想强调有声书带给听众的不同感受，则要在上述位置标注[Audiobook]。——译者注

❷ 根据 APA 格式第 7 版手册，网址前面的"Retrieved from/Accessed from"字样需要删除。只有包含了引用日期时才标注"Retrieved from"。——译者注

		续表
主编图书中的一章	参考文献	Board, J. (2016). The paradox of right and wrong. In R. Bolden, M. Witzel, & N. Linacre (Eds.), *Leadership paradoxes: Rethinking leadership for an uncertain world* (pp. 131–150). New York, NY: Routledge. ▶ 章节标题不用斜体、不加引号
	文内夹注	(Board, 2016)
期刊论文，单一作者	参考文献	Ivanov, M. (2016). Dynamic learning and strategic communication. *International Journal of Game Theory*, *45*, 627–653. https://doi.org/10.1007/s00182-015-0474-x Massaro, D. W. (2017). Reading aloud to children: Benefits and implications for acquiring literacy before schooling begins. *The American Journal of Psychology*, *130*, 63–72. https://doi.org/10.5406/amerjpsyc.130.1.0063 Wettersten, J. (2014). New social tasks for cognitive psychology; or, new cognitive tasks for social psychology [Abstract]. *American Journal of Pyschology*, *127*, 403. http://www.jstor.org/stable/10.5406/amerjpsyc.127.4.0403 ▶ 论文标题不用斜体、不加引号 ▶ 期刊卷号数字用斜体，期数、页码不用斜体。*Volume* 或 Vol.等词不出现 ▶ 如果期刊页码是全年连续编号的，就没必要注明期数。但是如果每一期都从第 1 页重新开始，就需要写出期数或月份，以便读者查询：*45*(2), 15–30. ▶ 引用文章的摘要时，在论文标题后面加方括号，标注"Abstract"字样 ▶ 即使引用的是印刷版文献，只要能找到 DOI 编码，都建议标注 DOI。如果只能提供网址，需要写清期刊的主页或所在数据库 Mitchell, T. (2002). McJihad: Islam in the U.S. global order. *Social Text*, *20*(4), 1–18. https://doi.org/10.1215/01642472-20-4_73-1 ▶ 括号内的数字 4 代表期数，不用斜体。由于每一期从第 1 页重新编号，因此期数 4 是必要的
	文内夹注	(Ivanov, 2016) (Massaro, 2017) (Wettersten, 2014) (Mitchell, 2002)
期刊论文，多位作者	参考文献	Lanis, R., Richardson, G., & Taylor, G. (2017). Board of director gender and corporate tax aggressiveness: An empirical analysis. *Journal of Business Ethics*, *144*, 577–596. https://doi.org/10.1007/s10551-015-2815-x Guo, S., Chen, D., Zhou, D., Sun, H., Wu, G., Haile, C., ... Zhang, X. (2007). Association of functional catechol O-methyl transferase (COMT) Val108Met polymorphism with smoking severity and age of smoking initiation in Chinese male smokers. *Psychopharmacology*, *190*, 449–456. https://doi.org/10.1007/s00213-006-0628-4 ▶ 最多可列出七位作者的名字；如果是八位或以上，列出前六位和最后一位，中间使用省略号隔开
	文内夹注	首次引用(Lanis, Richardson, & Taylor, 2017) 后续引用(Lanis et al., 2017) ▶ 如果作者人数是三至五人，首次引用要全部列出，后续引用可以只写第一位作者，然后加上"et al." (Guo et al., 2007) ▶ 如果作者人数为六位或以上，则全写成第一位作者加"et al."的形式。比如在文中第一次引用 Guo 的文章时，可以写： In their study of Chinese male smokers, Guo et al. (2007) find an association ...

期刊论文，非英语	参考文献	Maignan, I., & Swaen, V. (2004). La responsabilité sociale d'une organisation: Intégration des perspectives marketing et managériale. *Revue Française du Marketing*, 200, 51–66. ▶ 或 Maignan, I., & Swaen, V. (2004). La responsabilité sociale d'une organisation: Intégration des perspectives marketing et managériale [The social responsibility of an organization: Integration of marketing and managerial perspectives]. *Revue Française du Marketing*, 200, 51–66.
	文内夹注	(Maignan & Swaen, 2004)
报纸或杂志的文章，有作者	参考文献	Wingfield, N. (2017, September 11). The robots of Amazon. *The New York Times* (New York ed.), p. B1. ▶ APA 格式要求为报纸和杂志名称的开头添加 The ▶ 标注报纸页码时，要出现"p."或"pp."字样 Tsukayama, H. (2017, September 11). The iPhone is 10. Where does Apple go from here? *The Chicago Tribune*. Retrieved from www.chicagotribune.com Pandey, S. (2007, February 11). I read the news today, oh boy. *The Los Angeles Times* (Home ed.), p. M6. Retrieved from http://www.proquest.com ▶ 列出报纸、杂志或数据库的主页网址，而不是写特定文章的链接
	文内夹注	(Wingfield, 2017) 或必要时可以写作(Wingfield, September 11, 2017) (Tsukayama, 2017)或(Tsukayama, 2017, September 11) (Pandey, 2007)或(Pandey, 2007, February 11)
报纸或杂志的文章，无作者	参考文献	Retired U.S. general is focus of inquiry over Iran leak. (2013, June 28). *The New York Times* (New York ed.), p. A18. America and China talk climate change: Heating up or cooling down? (2009, June 11). *The Economist 391*(8635), 61. ▶ APA 格式对待杂志和期刊文章的方式比较类似。但是如果找不到杂志的卷数、期数信息，可以不写，直接写"p."或"pp."，后面标注页码范围即可
	文内夹注	("Retired U.S. general," 2013) ("Climate change," 2009)
书评	参考文献	Pugh, A. J. (2014, May). [Review of the book *The gender trap: Parents and the pitfalls of raising boys and girls*, by E. W. Kane]. *American Journal of Sociology* 119, 1773–1775. Vimercati, G. (2017, August 20). Soviet pseudoscience: The history of mind control [Review of the book *Homo Sovieticus: Brain waves, mind control, and telepathic destiny*, by W. Velminski]. *Los Angeles Review of Books*. Retrieved from https://lareviewofbooks.org
	文内夹注	(Pugh, 2014) (Vimercati, 2017)
未发表的论文、墙报论文、毕业论文或学位论文	参考文献	Tang, S. (2017, February). *Profit driven team grouping in social networks*. Paper presented at the Thirty-First AAAI Conference on Artificial Intelligence, San Francisco. ▶ 只需标注论文的年份、月份信息 Tomz, M., & Van Houweling, R. P. (2009, August). *Candidate inconsistency and*

续表

未发表的论文、墙报论文、毕业论文或学位论文	参考文献	*voter choice*. Unpublished manuscript, Stanford University and University of California, Berkeley. Retrieved from http://www.stanford.edu/~tomz/working/TomzVanHouweling-2009-08.pdf ▶ 文章的类型还可以写"Manuscript submitted for publication"或"Manuscript in preparation"等❶ Noble, L. (2006). *One goal, multiple strategies: Engagement in Sino-American WTO accession negotiations*. Unpublished master's thesis, University of British Columbia, Vancouver, Canada. ▶ 凡是无法从商业数据库访问、获取的毕业论文或学位论文，在 APA 格式中都视为未发表 Sierra, L. M. (2013). *Indigenous neighborhood residents in the urbanization of La Paz, Bolivia, 1910–1950* (Doctoral dissertation, State University of New York at Binghamton). Retrieved from ProQuest Dissertations and Theses (UMI No. 3612828). ▶ 结尾处的一串字母、数字组合是论文的登记号，方便读者在数据库中查找 ▶ 如果是能够从商业数据库检索到的学位论文，APA 格式对校名或机构名称就不做要求。但在我看来，写上去也无妨
	文内夹注	(Tang, 2017) (Tomz & Van Houweling, 2009) (Noble, 2006) (Sierra, 2013)
预印本	参考文献	Piatti, A. E. (2017). *On the extended stellar structure around NGC 288* [Preprint]. Retrieved from https://arxiv.org/pdf/1709.07284v1.pdf ▶ arXiv 是一个科技类的电子文献库，其中一些文章得以正式发表，但有些尚未发表。APA 格式建议你在即将完成论文时更新参考文献的信息；如果你引用的预印本后来已经出版了，请引用已出版的期刊文章 ▶ 此处标题为斜体，原因是文章尚未被期刊收录，视为一篇独立的作品 ▶ 一些期刊会将文章提前在线出版，此时你有两种引用方式（当然，论文临近发表时，你仍要对参考文献的状态进行更新） Ravigné, V., Dieckmann, U., & Olivieri, I. (2009). Live where you thrive: Joint evolution of habitat choice and local adaptation facilitates specialization and promotes diversity. *The American Naturalist*. Advance online publication. https://doi.org/10.1086/605369 ▶ 或 Ravigné, V., Dieckmann, U., & Olivieri, I. (in press). Live where you thrive: Joint evolution of habitat choice and local adaptation facilitates specialization and promotes diversity. *The American Naturalist*, *174*, E141–E169. https://doi.org/10.1086/605369
	文内夹注	(Piatti, 2017) (Ravigné, Dieckmann, & Olivieri, 2009)或 (Ravigné, Dieckmann, & Olivieri, in press) ▶ 后续引注可以写作(Ravigné et al., 2009)或(Ravigné et al., in press)

---

❶ 请对照参考 APA 格式第 7 版手册中提供的案例：
Lippincott, T., & Poindexter, E. K. (2019). *Emotion recognition as a function of facial cues: Implications for practice* [Manuscript submitted for publication]. Department of Psychology, University of Washington.——译者注

研究或技术报告，政策文件，其他灰色文献	参考文献	Environmental Protection Agency. (2016, December). *State of the science white paper: A summary of literature on the chemical toxicity of plastics pollution to aquatic life and aquatic-dependent wildlife* [White paper]. Retrieved from https://www.epa.gov/sites/production/files/2016-12/documents/plastics-aquatic-life-report.pdf Tarullo, D. (2017, October). *Monetary policy without a working theory of inflation*. [Working paper]. Retrieved from Hutchins Center on Fiscal & Monetary Policy, Brookings Institute website: https://www.brookings.edu/research/monetary-policy-without-a-working-theory-of-inflation/
	文内夹注	(Environmental Protection Agency, 2016) (Tarullo, 2017)
百科全书、词典	参考文献	Balkans: History. (1987). In *Encyclopaedia Britannica* (15th ed., Vol. 14, pp. 570–588). Chicago, IL: Encyclopaedia Britannica. Balkans. (2017). *Encyclopaedia Britannica online*. Retrieved from https://www.britannica.com/place/Balkans App. (2016, April 6). *Merriam-Webster*. Retrieved from http://www.merriam-webster.com/dictionary/app Protest, v. (1971). *Compact edition of the Oxford English dictionary* (Vol. 2, p. 2335). Oxford, England: Oxford University Press. ▶ "protest" 既是名词，也是动词。这里引用的是动词义项 Graham, G. (2015). Behaviorism. In E. N. Zalta (Ed.), *The Stanford encyclopedia of philosophy*. Retrieved from http://plato.stanford.edu/entries/behaviorism/ Emotion. (n.d.). Wikipedia. Retrieved November 8, 2017, from http://en.wikipedia.org/wiki/Emotion ▶ 尽管在 APA 格式中，大部分情况下你无需标注访问时间，但是无明确发表时间而且频繁更新的文章除外，所以上例中 Wikipedia 的文章就需要写清访问时间
	文内夹注	(Balkans: History, 1987) ▶ 你可能会想要插入"History"这个副标题，上例的做法能帮助读者辨别引文的具体来源，尤其是原文主体内容过长或同时存在多篇标题、时间都类似的参考文献的时候 (Balkans, 2017) (App, 2016) (Protest, 1971) (Graham, 2015) (Emotion, n.d.)
《圣经》《古兰经》	参考文献	▶ 无需列入参考文献，除非是为了指向特定的版本 *The five books of Moses: A translation with commentary.* (2004). Robert Alter (Trans. & Ed.). New York, NY: Norton.
	文内夹注	Deut. 1:2 (New Revised Standard Version). ▶ 第一次文内引用需要注明版本，后续的引用可省略版本信息 Gen. 1:1, 1:3–5, 2:4. ▶ 《圣经》的书目可以用 Exod.（出埃及记）、Lev.（利未记）、Num.（民数记）Cor.（歌林多书）等缩写形式 Qur'an 18:65–82.

古典著作	参考文献	▶ 无需列入参考文献，除非是为了指向特定的版本 Plato. (2006). *The republic*. R. E. Allen (Trans.). New Haven, CT: Yale University Press. (Original work, approximately 360 BC) Virgil. (2017). *The Aeneid*. D. Ferry (Trans.). Chicago, IL: University of Chicago Press. (Original work, approximately 19–29 BC)
	文内夹注	(Plato, trans. 2006) (Virgil, trans. 2017)
演讲、学术报告或课程讲座	参考文献	Wingfield, A. M. H. (2016, August 21). Presidential address at the annual meeting of the American Sociological Association, Seattle, WA. Rector, N. (2017, March 6). Course lecture at the University of Toronto, Toronto, Ontario, Canada. Shiller, R.J. (2011). Risk and financial crisis. [recorded lecture]. Retrieved from http://oyc.yale.edu/economics/econ-252-11/lecture-2
	文内夹注	(Wingfield, 2016) (Rector, 2017) (Shiller, 2011)
私人通信或访谈	参考文献	Gates, B. (2014, March 13). Bill Gates: The *Rolling Stone* interview. (J. Goodell, Interviewer). Retrieved from http://www.rollingstone.com/culture/news/bill-gates-the-rolling-stone-interview-20140313 Smith, H. (1941). Interview by J. H. Faulk [Audio file]. Library of Congress, Archive of Folk Culture, American Folklife Center, Washington, DC. Retrieved from http://hdl.loc.gov/loc.afc/afc9999001.5499a
	文内夹注	(Gates, 2014) (Smith, 1941) (J. M. Coetzee, personal interview, May 5, 2017) (D. A. Grossberg, personal communication, January 1, 2015) (anonymous US soldier, recently returned from Afghanistan, interviewed by author, August 22, 2010) ▶ 只有已经发表或存档的访谈才出现在参考文献列表中（上例中 Gates 和 Smith 的两段访谈）。如果是私人的谈话、传真、信件或访谈，其他人无从获取，也就只需在正文内部引用，上例中的 Coetzee、Grossberg 以及匿名美国士兵都属于这种情况——文内夹注需要完整地描述资料来源，包括详细的日期。不要将与研究参与者进行的访谈引用进来，相关内容应予以保密
电视节目	参考文献	Benioff, D., & Weiss, D. B. (Writers), & Graves, A. (Director). (2014). The children [Television series episode]. In D. Benioff & D. Weiss (Executive Producers) *Game of Thrones*. New York, NY: HBO. Carlock, R. (Writer), & Miller, M. B. (Director). (2009). Into the crevasse [Television series episode]. In T. Fey, L. Michaels, M. Klein, D. Miner, & R. Carlock (Producers), *30 Rock*. NBC. Retrieved from http://www.hulu.com/30-rock
	文内夹注	(Benioff, Weiss, & Graves, 2014) (Carlock & Miller, 2009)
电影	参考文献	Wallis, H. B. (Producer), & Huston, J. (Director/Writer). (1941). *The Maltese falcon* [Motion picture]. H. Bogart, M. Astor, P. Lorre, S. Greenstreet, E. Cook Jr. (Performers). Based on novel by D. Hammett. Warner Studios. United States: Warner

电影	参考文献	Home Video, DVD. (2000) Goolsby, K., & York, A. (Directors). (2015). *Tig* [Motion picture]. United States: Beachside Films and Netflix. Retrieved from www.netflix.com ▶ 必须包含的元素为：影片名称、导演、电影公司以及发行年份 ▶ 可以灵活处理的是：演员、制片人、编剧、剪辑、摄像等人物信息。根据实际写作需求决定是否列出，并按照相对于论文的重要性进行排序，把涉及的姓名写在片名和发行商之间 ▶ 如果在线访问的影片来自付费专区，或很容易能在网站（如 Netflix 或 Hulu 等）搜索到，可以只写网站主页地址
	文内夹注	(Wallis & Huston, 1941/2000) (Goolsby & York, 2015)
照片	参考文献	Adams, A. (1927). *Monolith, the face of Half Dome, Yosemite National Park* [Photograph]. Art Institute, Chicago. Adams, A. (1927). *Monolith, the face of Half Dome, Yosemite National Park* [Photograph]. Art Institute, Chicago. Retrieved from http://www.hctc.commnet.edu/artmuseum/anseladams/details/pdf/monlith.pdf
	文内夹注	(Adams, 1927)
图：地图、图表、曲线图或表格	图、表来源，其他注释	▶ 地图、图表、曲线图或表格的来源通常写在图表下方，而不是在正文中引注 *Note.* 2016 House election results map. *Politico.* Retrieved from http://www.politico.com/mapdata-2016/2016-election/results/map/house/ *Note.* From M. E. J. Newman (n.d.), Maps of the 2016 US presidential election. Retrieved from http://www-personal.umich.edu/~mejn/election/2016 *Note.* K. Smith, Urban Institute, (2013), Distribution of family income [Graph], 1963–2013. Retrieved from http://apps.urban.org/features/wealth-inequality-charts/ *Note.* From K. Menkhaus (2006/2007), Governance without government in Somalia: Spoilers, state building, and the politics of coping. *International Security, 31* (Winter), 79, fig. 1. ▶ 为地图、图表、曲线图或表格写一个描述性的标题，目的是让读者不看正文也能迅速了解图表内容 *Note.* All figures are rounded to nearest percentile. ▶ 这是一段常见的表格数据说明 **p<.05 **p<.01. Both are two-tailed tests. ▶ 这是有关统计表的概率说明
	参考文献	House election results map. (2016). *Politico.* Retrieved from http://www.politico.com/mapdata-2016/2016-election/results/map/house/ Newman, M. E. J. (n.d.). Maps of the 2016 US presidential election. Retrieved from http://www-personal.umich.edu/~mejn/election/2016 K. Smith, Urban Institute. (2013) Distribution of family income [Graph]. 1963-2013. Retrieved from http://apps.urban.org/features/wealth-inequality-charts/ Menkhaus, K. (2006/2007). Governance without government in Somalia: Spoilers, state building, and the politics of coping. *International Security, 31* (Winter), 74–106. https://doi.org/10.1162/isec.2007.31.3.74
	文内夹注	(2016 House election results, 2016) (Newman, n.d.) (K. Smith, 2013) (Menkhaus, 2006/2007)

续表

音乐唱片	参考文献	Johnson, R. (1998). Last fair deal gone down. On *Robert Johnson: King of the Delta blues singers* (Exp. ed.) [CD]. Columbia/Legacy. (Originally recorded 1936) ▶ "Exp. ed." 表示扩展版（expanded edition） Beyoncé. (2016). Formation. On *Lemonade*. Columbia / Parkwood Entertainment. Retrieved from http://tidal.com/us
	文内夹注	(Johnson, 1936/1998, track 5) (Beyoncé, 2016, track 11)
广告	参考文献	Advertisement for Letters from Iwo Jima [Motion picture]. (2007, February 6). *The New York Times*, p. B4. Drink outside the Lines. (2017, June 15). [Advertisement for Vitamin Water]. *Rolling Stone*, 17. Tab cola. [ca. late 1960s]. Be a mindsticker [Television advertisement]. Retrieved from http://www.dailymotion.com/video/×2s3qi_1960s-tab-commercial-be-a-mindstick_ads ▶ 如果不是准确日期，而是估计日期，则应使用方括号
	文内夹注	(advertisement *for Letters from Iwo Jima*, 2007) (Vitamin Water, 2017) (Tab cola [ca. late 1960s])
政府文献	参考文献	*A bill to promote the national security by providing for a national defense establishment: Hearings on S. 758 before the Committee on Armed Service, Senate,* 80th Cong. 1 (1947). ▶ "80th Cong. 1" 指的是第 1 页（不是第一次会议）。如果引用的是具体某个人的证词，可以标注在日期之后，即：(1947) (testimony of Gen. George Marshall). ▶ 若文件来自政府印刷局，要写全称"Government Printing Office"，不要用缩写（GPO） National Institute on Aging. (2015). *Caring for a person with Alzheimer's disease: Your easy-to-use guide from the National Institute on Aging.* Washington, DC: US Government Printing Office. ▶ 或 National Institute on Aging. (2015). *Caring for a person with Alzheimer's disease: Your easy-to-use guide from the National Institute on Aging.* Retrieved from https://permanent.access.gpo.gov/gpo66296/caring-for-a-person-with-alzheimers-disease.pdf Public Safety Canada (2017, September 24). Minister Goodale pays tribute to fallen police and peace officers. Retrieved from https://www.canada.ca/en/public-safety-canada/news/2017/09/minister_goodalepaystributetofallenpoliceandpeaceofficers.html Federal Bureau of Investigation [FBI]. (1972). *Investigation of John Lennon* (248 pages). Retrieved from http://foia.fbi.gov/foiaindex/lennon.htm
	文内夹注	(*Bill to promote national security*, 1947) (National Institute on Aging, 2015) (Public Safety Canada, 2017) (FBI, 1972)

数据库或数据集	参考文献	Maryland Department of Assessments and Taxation. (2017). *Real property data search*. Retrieved from https://sdat.dat.maryland.gov/RealProperty/Pages/default.aspx US Copyright Office. (2017). *Search copyright records*. Retrieved from http://www.copyright.gov/records/ Gleditsch, K. S., & Chiozza, G. (2009). *Archigos: A data base on leaders, 1875–2004* (Version 2.9). Retrieved from http://mail.rochester.edu/%7Ehgoemans/data.htm Pew Research Center. (2014). *Religious landscape study*. Retrieved from http://www.pewforum.org/religious-landscape-study/ United Nations Treaty Collection. (2010). *Databases*. Retrieved from http://treaties.un.org/ ▶ 如需指明上例数据库中的具体条目，可以写成： Convention on the prevention and punishment of the crime of genocide. (1948). In United Nations Treaty Collection, *Databases*. Retrieved from https://treaties.un.org/doc/Treaties/1951/01/19510112%2008-12%20PM/Ch_IV_1p.pdf
	文内夹注	(Maryland Department of Assessments, 2017) (US Copyright Office, 2017) (Gleditsch & Chiozza, 2009) (United Nations Treaty Collection, 2010) (Convention on the prevention and punishment of the crime of genocide, 1948)
诊断性测试	参考文献	*MMPI-2: Restructured clinical (RC) scales*. (2003). Minneapolis: University of Minnesota Press. Ben-Porath, Y. S., & Tellegen, A. (2008). *MMPI-2-RF (Minnesota Multiphasic Personality Inventory-2): Manual for administration, scoring, and interpretation*. Minneapolis: University of Minnesota Press. ▶ 上例引用的是测试的实施手册 Ben-Porath, Y. S. (2012). *Interpreting the MMPI-2-RF*. Minneapolis: University of Minnesota Press. ▶ 上例是测试的解释手册 Q Local (Version 3.3) [Computer software]. (2017). Minneapolis, MN: Pearson Assessments. ▶ 上例引用的是测试的计分软件
	文内夹注	(*MMPI-2 RC scales*, 2003) (Ben-Porath & Tellegen, 2008) (Ben-Porath, 2012) (Q Local, 2017)
诊断手册	参考文献	Lingiardi, V., & McWilliams, M. (Eds.). (2017). *Psychodynamic diagnostic manual* (2nd ed.). New York, NY: Guilford Press. American Psychiatric Association. (2000). *Diagnostic and statistical manual of mental disorders* (4th ed., text rev. [*DSM-IV-TR*]). Washington, DC: Author. ▶ 或 American Psychiatric Association. (2000). *Diagnostic and statistical manual of mental disorders* (4th ed., text rev. [*DSM-IV-TR*]). https://doi.org/10.1176/appi.books.9780890423349

诊断手册	文内夹注	(Lingiardi & McWilliams, 2017) 首次引用(American Psychiatric Association, *Diagnostic and statistical manual of mental disorders* [*DSM-IV-TR*], 2000) 后续引用，注意标题的斜体(*DSM-IV-TR*)
网站、网页	参考文献	American Historical Association [Homepage]. (2017). Retrieved August 1, 2017, from https://www.historians.org ▶ 许多网页都没有明确的作者，可以列出赞助该网站的组织或机构 Lipson, C. (2010). "Advice on getting a good recommendation." Retrieved from http://www.charleslipson.com/Getting-a-good-recommendation.htm *The Dick Van Dyke show*: Series summary. (n.d.). *Sitcoms online*. Retrieved from http://www.sitcomsonline.com/thedickvandykeshow.html ▶ 如果一个网站或网页没有显示它的版权或更新日期，则可在年份的括号中标注"n.d."，表示没有日期（no date） ▶ 通常在正文括号内提供网站的地址就足够了
	文内夹注	(American Historical Association, 2017) (Lipson, 2010) (*The Dick Van Dyke show*: Series summary, n.d.)
博客文章或评论	参考文献	Pearce, F. (2017, September 18). In a stunning turnaround, Britain moves to end the burning of coal. [Blog post]. *Yale Environment 360*. Retrieved from http://e360.yale.edu/features/in-a-stunning-turnaround-britain-moves-to-end-the-burning-of-coal ▶ 如果被引内容没有标题，则可写作： Pearce, F. (2017, September 18). Untitled. [Blog post]. *Yale Environment 360* ... Dinah. (2017, August 1). Inpatient psychiatry: Not all bad. [Blog post]. *Shrink Rap*. Retrieved from http://psychiatrist-blog.blogspot.com/ Catlover. (2017, August 5). Re: Inpatient psychiatry: Not all bad. [Blog comment]. *Shrink Rap*. Retrieved from https://www.blogger.com/comment.g?blogID=26666124&postID=498010513235225858 ▶ 若上例中的 Catlover 在同一天发布了多条评论，而且无法为具体某条评论提供链接，则需要在日期后面标注评论的时间，例如：(2017, August 5, 12:53 a.m.)
	文内夹注	(Pearce, 2017) (Dinah, 2017) (Catlover, 2017)
视频	参考文献	Archer Productions (Producer). (1951). *Duck and cover* [Video file]. Produced for the Federal Civil Defense Administration. Retrieved from https://en.wikipedia.org/wiki/File:DuckandC1951.ogv Jury convicts Martin Shkreli on three counts of security fraud. (2017, August 4). [Video file]. Retrieved from https://www.nbcnews.com/video/jury-convicts-martin-shkreli-on-three-counts-of-securities-fraud-1017719875664 Wang, T. (2016, September). The human insights missing from big data. [Video file]. Retrieved from https://www.ted.com/talks/tricia_wang_the_human_insights_missing_from_big_data Stevens, M. (2017, July 17). How much of the earth can you see at once? [Video file]. Retrieved from https://youtu.be/mxhxL1LzKww

续表

视频	参考文献	Windfall. (July 29, 2017). Re: How much of the earth can you see at once? [Comment on video blog post]. Retrieved from https://youtu.be/mxhxL1LzKww ▶ 若视频网址过长，且能在网站上搜索获取，则可以考虑只引用网站的主页，比如：Retrieved from https://www.ted.com/
	文内夹注	(Archer Productions, 1951) ("Jury convicts Martin Shkreli," 2017) (Wang, 2016) (Stevens, 2017) (Windfall, 2017)
多媒体应用和软件	参考文献	▶ APA 格式引用移动应用和电脑软件的方式相同 ▶ 像 Microsoft 或 Adobe 之类的标准软件无需写入参考文献，但一些专业软件或应用需要 EA Sports. (2017). Madden NFL 18 [Video game]. Retrieved from https://www.easports.com/madden-nfl Groundspeak. (2017). Geocaching. (v. 5.6.1) [Mobile application software]. Retrieved from https://www.geocaching.com/play Mathworks. (2014). Matlab. (versionR2014b) [Computer software]. Retrieved from https://www.mathworks.com/products/matlab.html
	文内夹注	(EA Sports, 2017) (Groundspeak, 2017) (Mathworks, 2014) (Microsoft home & business, 2016)
播客	参考文献	Burriss, R. (Producer). (2017, August 1). Conflict and reconciliation [Audio podcast]. In *The Psychology of Attractiveness*. Retrieved from http://psychologyofattractivenesspodcast.blogspot.com/2017/08/conflict-and-reconciliation-01-aug-2017.html ▶ 或，为了避免网址过长： Burriss, R. (Producer). (2017, August 1). Conflict and reconciliation [Audio podcast]. In *The Psychology of Attractiveness*. Retrieved from http://psychologyofattractivenesspodcast.blogspot.com Shipping container grow rooms, Uber for dogs, and plutonium on Pluto. (2015, July 27). [Video podcast]. In *Geekbeat*. Retrieved from http://geekbeat.tv/shipping-container-grow-rooms-uber-for-dogs-and-plutonium-on-pluto/ ▶ 或 Draney, G. (Host). (2015, July 27). Shipping container grow rooms, Uber for dogs, and plutonium on Pluto. [Video podcast]. In Geekbeat. Retrieved from http://geekbeat.tv
	文内夹注	(Burris, 2017) ("Shipping container," 2015) ▶ 或 (Draney, 2015)

社交媒体 (Facebook, Instagram, Twitter)	参考文献	Simpson, M. [Maggie]. (n.d.) [Profile]. Retrieved from http://www.facebook.com/maggiesimpson ▶ 假设另一条参考文献的作者是 Marge Simpson，此时仅凭首字母 M 可能不足以区分不同的作者，那么需要仿照上例，将名字备注在方括号内。文内引用时，应写出完整的姓名 Grammar Girl. (2017, October 5). Diminutives often express smallness [Facebook status update]. Retrieved from https://www.facebook.com/GrammarGirl/posts/10157671243605228 Tyson, N. D. [neiltyson]. (2017, July 7, 1:25pm). Always seemed to me that millipedes have more legs than are necessary [Tweet]. Retrieved from https://twitter.com/neiltyson/status/883421859072458752 ▶ 如果可能的话，提供信息或页面的存档地址（"永久链接"）。在 Facebook 和 Twitter 等网站上，可以通过日期和时间戳获取此类网址 Richards, C. [coryrichards]. (2017, May 4). A quiet if not solemn day on Everest [Instagram photo]. Retrieved from https://www.instagram.com/p/BUekBZLgDOW/
	文内夹注	(Maggie Simpson, n.d.) (Grammar Girl, 2017) (Tyson, 2017) (Richards, 2017)
邮件、短信（私信）	参考文献	▶ 私人邮件、短信或其他类型的私聊信息无法被第三方检索，无需写入参考文献。如果是能通过网址获取的条目，引用格式为： Abrams, J. (2015, April 28). Fw: a Favor. [Email]. Retrieved from https://assets.documentcloud.org/documents/3516674/Abrams amp-Bank-Phoenix-Electric.pdf
	文内夹注	(Abrams, 2015) ▶ 鉴于私人邮件和即时消息不写入参考文献，文内引用时应该尽量提供完整描述 (E. Leis, email message to author, 2018, May 3) (M. H. Lipson, instant message to J. S. Lipson, 2015, March 9) ▶ 为区分不同消息或体现某条信息的重要性，可以标注具体时间 (M. H. Lipson, instant message to J. S. Lipson, 2015, March 9, 11:23 a.m.)
电子论坛、邮件列表	参考文献	▶ 未存档的邮件列表或讨论组的信息无法被第三方检索，因此无需包含在参考文献列表内。若是能通过网址获取的条目，则要写入参考文献 Ngo, Q. (2017, July 14). Re: what are some unknown truths about college life? [Online forum comment]. Retrieved from https://www.quora.com/What-are-some-unknown-truths-about-college-life ▶ 若知道用户的真实姓名，建议注明，网名则放入方括号 ▶ 若无法从网址看出论坛或邮件列表的名称，应在参考文献条目内写出论坛（或邮件列表）的名称
	文内夹注	(Ngo, 2017)

APA 格式下的参考文献不允许出现太多缩写。常见的缩写形式也是大小写兼有（表8.3）。原因不明，我们只需照做。

表8.3　APA格式：常见的缩写

全称	缩写	全称	缩写
chapter	chap.	part	Pt.
edition	ed.	revised edition	Rev. ed.
editor	Ed.	second edition	2nd ed.
no date	n.d.	supplement	Suppl.
number	No.	translated by	Trans.
page	p.	volume	Vol.（例如：Vols. 2–5）
pages	pp.	volumes	vols.（例如：3 vols.）

9

CSE 格式：生物科学

9.1 CSE 格式的三种引注方式

国际科学编辑委员会（Council of Science Editors）制定的 CSE 格式广泛用于生命科学领域的研究论文、期刊和图书。这套格式是基于美国国家医学图书馆所采用的国际原则设计而成。详细信息可以参考芝加哥大学出版社 2014 年出版的 *Scientific Style and Format: The CSE Manual for Authors, Editors, and Publishers*（第 8 版），或访问 www.scientificstyleandformat.org 获取。

实际应用时，CSE 格式允许使用下列三种引注方法：

● 顺序编码制：引文根据在正文出现的顺序，依次编号为（1）、（2）、（3）……，论文最后完整的参考文献列表也用同样的次序出现，无需考虑字母顺序。

● 姓名编码制：文末的参考文献按作者姓氏字母顺序排列，文章中的引文按照对应的数字进行编号。这就意味着，如果文中第一条引文属于第 8 条文献，就应该编号为（8），而且它在文章的任何位置再次出现都对应数字（8）。

● 著者-出版年制：文内夹注的形式是作者加年份，如（McClintock 2017），文末完整的参考文献列表按字母顺序排列。这与 APA 格式类似。

一旦你选定了某种方法，就要在文章内部保持一致。写作前，建议询问导师的要求和偏好。

顺序编码制：文章里用到的首篇参考文献标记为 1，第二篇为 2，以此类推。标记的格式可以是方括号[1]、上标 [1] 或圆括号（1）。在论文的最后，从第一个引用的条目开始列出所有文献，无需考虑字母的先后顺序。若你的第一条引文来自 Zangwill 教授，那么参考文献列表就从 Zangwill 开始。即使你的最后一次引用又涉及这篇文献，它的编号依然是[1]。如果你希望同时引用多个条目，可以将数字一次性列出，中间用逗号隔开，写作：[1,3,9]、[1,3,9]或（1,3,9）。如果包含了连续的数字，可以使用连字符，即：4–6,12–18。

姓名编码制：首先形成的是文末按字母顺序排列的参考文献列表，然后为文献依次编号，这样一来，每条文献所包含的引文也就有了自己的编号。由于字母 Z 次序靠后，假设 Zangwill 的文章排在第 36 位，那么每次文内引注都要按这个数字标记，哪怕你的第一条引文就用到了这篇文章；而第二条引文可能标为[23]，下一条对应的文献也许是[12]。文内引注仍然是上标、方括号、圆括号三种格式。如果一次性引用多篇文献，可以把所有数字列出来：^{4,15,22}、[4,15,22]或（4,15,22）。如果涉及连续的数字，就写成（1–3）。最后，比较复杂的情况或许是（4,16–18,22）。

著者-出版年制：文内夹注需用（作者-年份）的格式，中间没有逗号，例如（Cronin and Siegler 2016）以及（Siegler et al. 2017）。将所有条目写入参考文献，并按文献作者姓氏的首字母排序。如果涉及同一作者的不同文章，按发表时间由远及近的顺序排列。如果是同一作者、相同年份的不同文章，处理方法也是如此，文献列表按发表的月份排序，在文中标记为 2017a 和 2017b。比如，文中引用了 Susan Lindquist 的许多作品，夹注可能是（Lindquist 2013d, 2014a, 2017h），对应的三篇文献都需要列在参考文献列表中。

同理，若在一处引用不同作者的多篇作品，可以一并列出，用分号隔开，例如：（Ma and Lindquist 2013; Outeiro and Lindquist 2015; Liebman 2017）。

倘若作者姓名已在正文中提及，则无需在夹注的括号中重复，比如：According to LaBarbera（2010），this experiment …

如果 LaBarbera 有十位甚至十五位共同作者怎么办？在科技领域，这很正常——实验结果发表的背后，有时凝结了几十人的共同努力，每个人的名字都值得体现在成果中。我的同事 Henry Frisch 是一位高能物理学家，他的一篇文章有将近 800 位共同作者❶。我出生在一个小镇，坦白讲，这名单已经比镇上的电话号码本还要长了。

那么著者-出版年制允许你列出多少位呢？建议还是要把握尺度。如果你决定把前 700 位写进去，可能论文刚一开始就突破了字数限制，如此看来，写科技论文倒是轻松了许多。

事实上，CSE 格式的要求非常清楚，跟 700 个名字也有一些差距。如果原文献只有两位作者，则都应在夹注中列出，中间用单词"and"连接。如果有三位或更多作者，只写出第一位，后面用"et al."，例如（LaBarbera et al. 2010）。至于参考文献列表的制作格式，我们稍后再来讲解。CSE 格式快速对比见表 9.1。

表9.1　CSE 格式快速对比

引注方法	文内引用	文末参考文献
顺序编码制	(1), (2), (3), (4)	根据在文中出现的先后顺序排列
姓名编码制	(31), (2), (13), (7)	根据作者姓氏首字母排序
著者-出版年制	(Shapiro 2016)	根据作者姓氏首字母排序

---

❶ 然而 Frisch 教授的习惯是：只有参与了写作才将自己列为作者。他的做法并不常见，但许多科学家是赞同的。针对这一议题目前没有明确的规定，但是为了应对越来越长的作者名单，一些科学家开始倡议，要求共同作者具体说明为论文所做的贡献。Frisch 对此发表的评论，可以从他的网页上获取：http://hep.uchicago.edu/~frisch/authorship.pdf

9.2 参考文献列表的格式

三种引注方法都要求在文章最后列出所有文献。CSE 格式下的参考文献以简洁为主。无需写出作者完整的名字，用首字母代替即可，而且大写字母后面不需要英文句号和空格（例如：Stern HK）。

期刊的名称也有标准的缩写方式，比如 PubMed 期刊数据库中列出的缩写，或参考 http://www.ncbi.nlm.nih.gov/journals 上的相关信息。在 CSE 格式中，缩写的期刊名称中间不需要插入英文句号，*Journal of Bioscience and Bioengineering* 的对应缩写为 J Biosci Bioeng（没有句点）。

CSE 格式把参考文献的标题视为"句子"，所以只有"句首"的第一个字母、专有名词、冒号后第一个词的首字母需要大写。标题用普通字体，不用斜体。

如果你参考的是网上的资料，就按电子版引用，不要按纸质版文献引用，毕竟两个版本可能会有差别。为确保读者了解文献的来源，CSE 要求在电子版本的参考文献中提供：①文件的修改（或发表）日期和访问日期；②文件的网址或 DOI 编码，二者相比，更推荐使用 DOI（在前面加上 https://doi.org/）。请看一组对比：

引用纸质版文献	Pantazopoulou M, Boban M, Foisner R, Ljungdahl PO. 2016. Cdc48 and Ubx1 participate in a pathway associated with the inner nuclear membrane that governs Asi1 degradation. J Cell Sci. 129(20): 3770–3780.
引用线上版文献	Pantazopoulou M, Boban M, Foisner R, Ljungdahl PO. 2016. Cdc48 and Ubx1 participate in a pathway associated with the inner nuclear membrane that governs Asi1 degradation. J Cell Sci. [accessed 2017 Sep 14];129(20):3770–3780. https://doi.org/10.1242/jcs.189332.

与绝大多数印刷出版的文章一样，上例中的文章在发表后没有再度修改。但是预印本和电子期刊则不同，它们的修改比较频繁。但只要你按上述要求提供了相关信息，读者就能知道你引用的是哪个版本。修改日期写入方括号，置于访问日期之前。

修改过的文献	Pantazopoulou M, Boban M, Foisner R, Ljungdahl PO. 2016. Cdc48 and Ubx1 participate in a pathway associated with the inner nuclear membrane that governs Asi1 degradation. J Cell Sci. [modified 2016 Dec 2; accessed 2017 Sep 14];129(20):3770–3780. https://doi.org/ 10.1242/jcs.189332.

9.3 文内引注和参考文献列表的引用举例

刚刚提到的所有引用规则都无需背诵，因为细节实在是太多了。我还是按照惯例到表格中去解释它们，然后为你提供足够的例证。随着实践的增多，你会越来越熟悉 CSE 格式的各种注意事项。

下面的几张表格（表9.2~表9.5）包含了文内引注和参考文献（Reference list）的写法，而且 CSE 的三种方法都会讲到。并不是所有期刊都严格遵守这些体例，你在阅读文献时也许会发现一些变体。有些期刊可能对允许列出的作者数量有新的要求，比如可以写出前三位作者再加"et al."，或是可以写出前 26 位作者（在这里对第 27 位合著者表示遗憾），而 CSE 格式建议最多列出某条文献的十位作者，然后再加"et al."。

表9.2　CSE 格式：著者-出版年制

期刊论文	参考文献	Dannemann M, Kelso J. 2017. The contribution of Neanderthals to phenotypic variation in modern humans. Am J Hum Genet. 101(4):578–589. ▶ 如果期刊页码是全年连续编号的，CSE 格式就不要求注明期数。当然，写出来也没问题 Wong KK et al. 2007. A comprehensive analysis of common copy-number variations in the human genome. Am J Hum Genet. 80(1):91–104. ▶ Wong 的这篇论文共有十一位作者。CSE 格式允许写出前十位，然后标注"et al."，但特定的期刊可能有不同的要求。一些期刊允许全部列出，大多数期刊只允许写出前两位 Guerrero-Sanchez VM et al. 2017. Holm oak (*Queras ilex*) transcriptome. *De novo* sequencing and assembly analysis. Front Mol Biosci [modified 2017 Oct 6; accessed 2017 Oct 21];4:70. https://doi.org/10.3389/fmolb.2017.00070. ▶ 注意观察 CSE 格式独特的日期写法（2017 Oct 21）
	文内夹注	(Dannemann and Kelso 2017) (Wong et al. 2007) ▶ 如果你的论文还引用了 Wong 与其他人合作发表于 2007 年的作品，文内夹注就需要将共同作者也列上，提示读者此处引用的到底是哪篇文献，例如： (Wong, deLeeuw, et al. 2007) (Guerrero-Sanchez et al. 2017)
图书，单一作者	参考文献	Bickley LS. 2016. Bates' guide to physical examination and history taking. Alphen aan de Rijn (Netherlands): Wolters Kluwer. Vetter RS. 2015. The brown recluse spider. Ithaca (NY): Cornell University Press. ▶ 如果出版社位于知名城市，州的名称可以省略
	文内夹注	(Bickley 2016) (Vetter 2015) ▶ 若引用同一作者在不同年份发表的一系列作品： (Vetter 2015, 2016a, 2016b, 2018) ▶ 若作者的姓氏相同，而且作品的发表年份也相同，则应标注名字的首字母，以示区分 (Vetter RS 2015; Vetter T 2015)

续表

图书，多位作者	参考文献	Provost JJ, Colabroy KL, Kelly BS, Wallert MA. 2016. The science of cooking: Understanding the biology and chemistry behind food and cooking. Hoboken (NJ): John Wiley & Sons. ▶ 此处最多列出十位作者，然后加上"et al."
	文内夹注	(Provost et al. 2016) ▶ 假设只有两位作者，则均应列出，比如：(Provost and Colabroy 2016)
电子图书或在线图书	参考文献	Schmitz OJ. 2016. The new ecology: Rethinking a science for the Anthropocene. Princeton (NJ): Princeton University Press. [accessed 2017 Oct 22]. Google Play Books. ▶ 如果是在线图书，在条目最后应标注网址或 DOI 编码，替代上例中的设备或服务商信息
	文内夹注	(Schmitz 2016)
图书，多个版本	参考文献	Gerard JE. 2014. Principles of environmental chemistry. 3rd ed. Burlington (MA): Jones and Bartlett. Snell RS. 2007. Clinical anatomy by regions. 8th ed. Philadelphia: Lippincott Williams & Wilkins. ▶ 如果是修订版（revised edition），可以将版次"8th ed."替换为"Rev. ed."
	文内夹注	(Gerard 2014) (Snell 2007)
图书，多个版本，无作者	参考文献	Publication manual of the American Psychological Association. 2010. 6th ed. Washington (DC): American Psychological Association.
	文内夹注	(Publication manual ... 2010) ▶ 无需在作者姓名的位置标注 Anonymous。CSE 格式的做法是先写出标题的第一个词或前几个词，然后加上省略号，最后写上年份
主编图书	参考文献	Rish, I, Cecchi GA, Lozano A, NiculescuMizil A, editors. 2014. Practical applications of sparse modeling. Cambridge (MA): MIT Press.
	文内夹注	(Rish et al. 2014)
主编图书中的一章	参考文献	McGrath J. 2015. Environmental factors and gene-environment interactions. In: Mitchell KJ, editor. The genetics of neurodevelopmental disorders. Hoboken (NJ): John Wiley and Sons. p. 111–127.
	文内夹注	(McGrath 2015)
预印本	参考文献	Muir CD. 2017. Light and growth form interact to shape stomatal ratio among British angiosperms. Preprint. bioRxiv. [posted 2017 Jul 14, accessed 2017 Oct 22]. https://doi.org/10.1101/163873. ▶ CSE 格式的第 8 版官方手册没有明确规定如何引用预印本。此处提供的格式符合手册中的整体引用原则
	文内夹注	(Muir 2017)
政府文献，纸质或在线	参考文献	Marinopoulos SS, Dorman T, Ratanawongsa N, Wilson LM, Ashar BH, Magaziner JL, Miller RG, Thomas PA, Prokopowicz GP, Qayyum R, et al. 2007. Effectiveness of continuing medical education. Rockville (MD): Agency for Healthcare Research and Quality. AHRQ Pub. No. 07-E006.

续表

政府文献，纸质或在线	参考文献	[EPA] Environmental Protection Agency (US), Office of Air and Radiation, Indoor Environments Division. 2012. A brief guide to mold, moisture, and your home. Washington (DC): EPA. [modified 2012 Sep 1; accessed 2015 Nov 28]. https://www.epa.gov/nscep. ▶ 如果机构既是作者又是出版商，那么在出版商的位置可以对机构进行缩写
	文内夹注	(Marinopoulos et al. 2007) (EPA 2012)
视频	参考文献	3D human anatomy: Regional edition [DVD-ROM]. 2009. London (England): Primal Pictures. The secret social life of a solitary puma [video]. 2017 Oct 12, 1:43 minutes. Scientific American. [accessed 2017 Nov 19]. https://www.scientificamerican.com/video/the-secret-social-live-of-a-solitary-puma/.
	文内夹注	(3D human anatomy 2009) (Secret social life 2017)
数据库	参考文献	RCSB Protein Data Bank. 1971–. [accessed 2017 Aug 1]. http://www.rcsb.org/pdb/home/home.do. Swedish Life Sciences Database. [date unknown]–. Zurich (Switzerland): Venture Valuation. [accessed 2017 Sep 15]. http://www.swedishlifesciences.com/se/portal/index.php. ▶ 上例中的两个数据库都在不断更新，表示方法为：先写出创立时间（如果不知道，在方括号内写 date unknown），后面加一个短破折号，也可使用连字符加空格
	文内夹注	(RCSB Protein Data Bank 1971–) (Swedish Life Sciences Database [accessed 2017]) ▶ 如果没有其他日期，则列上访问日期
网站、网页或博客	参考文献	[CSE] Council of Science Editors. c2017. Wheat Ridge (CO): Council of Science Editors; [accessed 2017 Oct 19]. https://www.councilscienceeditors.org/resource-library/editorial-policies/cse-policies/. ▶ 如果可能，请注明发布日期或版权日期（如c2017） [USDA] US Department of Agriculture, Agricultural Research Service. Washington (DC): USDA; [modified 2017 Feb 2; accessed 2017 Nov 8]. https://www.ars.usda.gov/. Haran M. 2016 Sep 7. Active parenting behaviors of the Caribbean spiny lobster. BMC series blog. [accessed 2017 Jul 29]. https://blogs.biomedcentral.com/bmcseriesblog/2016/09/07/active-parenting-behaviors-caribbean-spiny-lobster. ▶ 若不能明显看出是一篇博客文章，要在日期后插入[blog]字样
	文内夹注	(CSE c2017) (USDA 2017) (Haran 2016)
播客	参考文献	Semple I. Tomorrow's technology: From asteroid mining to programmable matter. 2017 Nov 15, 30:42 minutes. The Guardian science weekly podcast. [accessed 2017 Nov 19]. https://www.theguardian.com/science/audio/2017/nov/15/tomorrows-technology-from-asteroid-mining-to-programmable-matter-science-weekly-podcast.
	文内夹注	(Semple 2017)

表 9.3 展示的是 CSE 顺序编码制和姓名编码制的参考文献格式。与表 9.2 的主要区别在于，日期出现在文献条目的尾部。两个表格用到的案例相同，方便你进行比较。

表9.3 CSE格式：顺序编码制和姓名编码制

期刊论文	参考文献	Dannemann M, Kelso J. The contribution of Neanderthals to phenotypic variation in modern humans. Am J Hum Genet. 2017;101(4):578–589. Wong KK, deLeeuw RJ, et al. A comprehensive analysis of common copy-number variations in the human genome. Am J Hum Genet. 2007;80(1): 91–104. Guerrero-Sanchez VM, et al. Holm oak (*Queras ilex*) transcriptome. *De novo* sequencing and assembly analysis. Front Mol Biosci. 2017 [modified 2017 Oct 6; accessed 2017 Oct 21];4:70. https://doi.org/10.3389/fmolb.2017.00070.
图书，单一作者	参考文献	Bickley LS. Bates' guide to physical examination and history taking. Alphen aan de Rijn (Netherlands): Wolters Kluwer; 2016. Vetter RS. The brown recluse spider. Ithaca (NY): Cornell University Press; 2015.
图书，多位作者	参考文献	Provost JJ, Colabroy KL, Kelly BS, Wallert MA. The science of cooking: Understanding the biology and chemistry behind food and cooking. Hoboken (NJ): John Wiley & Sons; 2016. ▶ 此处最多列出十位作者，然后加上"et al."
电子图书或在线图书	参考文献	Schmitz OJ. The new ecology: rethinking a science for the Anthropocene. Princeton (NJ): Princeton University Press; 2016 [accessed 2017 Oct 22]. Google Play Books. ▶ 如果是在线图书，在条目最后应标注网址或 DOI 编码
图书，多个版本	参考文献	Gerard JE. Principles of environmental chemistry. 3rd ed. Burlington (MA): Jones and Bartlett; 2014. Snell RS. Clinical anatomy by regions. 8th ed. Philadelphia: Lippincott Williams & Wilkins; 2007.
图书，多个版本，无作者	参考文献	Publication manual of the American Psychological Association. 6th ed. Washington (DC): American Psychological Association; 2010.
主编图书	参考文献	Rish, I, Cecchi GA, Lozano A, NiculescuMizil A, editors. Practical applications of sparse modeling. Cambridge (MA): MIT Press; 2014.
主编图书中的一章	参考文献	McGrath J. Environmental factors and gene-environment interactions. In: Mitchell KJ, editor. The genetics of neurodevelopmental disorders. Hoboken (NJ): John Wiley and Sons; 2015. p. 111–127.
预印本	参考文献	Muir CD. Light and growth form interact to shape stomatal ratio among British angiosperms. Preprint. bioRxiv; 2017 [posted 2017 Jul 14, accessed 2017 Oct 22]. https://doi.org/10.1101/163873.
政府文献	参考文献	Marinopoulos SS, Dorman T, Ratanawongsa N, Wilson LM, Ashar BH, Magaziner JL, Miller RG, Thomas PA, Prokopowicz GP, Qayyum R, Bass EB. Effectiveness of continuing medical education. Rockville (MD): Agency for Healthcare Research and Quality; 2007. AHRQ Pub. No. 07-E006. [EPA] Environmental Protection Agency (US), Office of Air and Radiation, Indoor Environments Division. A brief guide to mold, moisture, and your home. Washington (DC): EPA; 2012 [modified 2012 Sep 1; accessed 2015 Nov 28]. https://www.epa.gov/nscep

续表

视频	参考文献	3D human anatomy: Regional edition [DVD-ROM]. London (England): Primal Pictures; 2009. The secret social life of a solitary puma [video]. Scientific American. 2017 Oct 12, 1:43. [accessed 2017 Nov 19]. https://www.scientificamerican.com/video/the-secret-social-live-of-a-solitary-puma/.
数据库	参考文献	RCSB Protein Data Bank. 1971–[accessed 2017 Aug 1]. http://www.rcsb.org/pdb/home/home.do. Swedish Life Sciences Database. Zurich (Switzerland): Venture Valuation. [date unknown]–[accessed 2017 Sep 15]. http://www.swedishlifesciences.com/se/portal/index.php.
网站、网页或博客	参考文献	[CSE] Council of Science Editors. Wheat Ridge (CO): Council of Science Editors; c2017 [accessed 2017 Oct 19]. https://www.councilscienceeditors.org/resource-library/editorial-policies/cse-policies/. [USDA] US Department of Agriculture, Agricultural Research Service. Washington (DC): USDA; [modified 2017 Feb 2; accessed 2017 Nov 8]. https://www.ars.usda.gov/. Haran M. Active parenting behaviors of the Caribbean spiny lobster. BMC series blog. 2016 Sep 7 [accessed 2017 Jul 29]. https://blogs.biomedcentral.com/bmcseriesblog/2016/09/07/active-parenting-behaviors-caribbean-spiny-lobster.
播客	参考文献	Semple I. Tomorrow's technology: From asteroid mining to programmable matter. The Guardian science weekly podcast. 2017 Nov 15, 30:42 minutes [accessed 2017 Nov 19]. https://www.theguardian.com/science/audio/2017/nov/15/tomorrows-technology-from-asteroid-mining-to-programmable-matter-science-weekly-podcast.

如表 9.3 所示，顺序编码制和姓名编码制的每一条参考文献都采用相同的格式，但完整的文献列表是有区别的。

你需要注意文献的排序：

- 姓名编码制：按照作者姓氏的首字母排序。
- 顺序编码制：按照引文在文中出现的先后顺序排列。

为了让你看得更加直观，我们选取了一篇论文的第一句话，请对比两种体系（表 9.4、表 9.5）下，引注方式和文献列表的不同之处。

表 9.4　CSE 格式：顺序编码制（参考文献排序举例）

论文开场白	This research deals with the ABC transporter family and builds on prior studies by Nguyen et al.,[1] Sheps et al.,[2] and Kerr.[3]
参考文献 （取决于在文中出现的次序）	1. Nguyen VNT, Moon S, Jung, K. Molecular biology: Genome-wide expression analysis of rice ABC transporter family across spatio-temporal samples and in response to abiotic stresses. J Plant Physiol. 2014;171(14):1276–1288.

续表

参考文献 （取决于在文中出现的次序）	2. Sheps JA, Ralph S, Zhao Z, Baillie DL, Ling V. The ABC transporter gene family of Caenorhabditis elegans has implications for the evolutionary dynamics of multidrug resistance in eukaryotes. Genome Biol. 2004;5(3):R15. 3. Kerr ID. Sequence analysis of twin ATP binding cassette proteins involved in translational control, antibiotic resistance, and ribonuclease L inhibition. Biochem Biophys Res Commun. 2004;315(1):166–173. ▶ Nguyen 的文章最先被引用，因此排在首位

表9.5　CSE格式：姓名编码制（参考文献排序举例）

论文开场白	This research deals with the ABC transporter family and builds on prior studies by Nguyen et al.,[2] Sheps et al.,[3] and Kerr.[1]
参考文献 （字母顺序）	1. Kerr ID. Sequence analysis of twin ATP binding cassette proteins involved in translational control, antibiotic resistance, and ribonuclease L inhibition. Biochem Biophys Res Commun. 2004;315(1):166–173. 2. Nguyen VNT, Moon S, Jung, K. Molecular biology: Genome-wide expression analysis of rice ABC transporter family across spatio-temporal samples and in response to abiotic stresses. J Plant Physiol. 2014;171(14):1276–1288. 3. Sheps JA, Ralph S, Zhao Z, Baillie DL, Ling V. The ABC transporter gene family of Caenorhabditis elegans has implications for the evolutionary dynamics of multidrug resistance in eukaryotes. Genome Biol. 2004;5(3):R15. ▶ 按照姓氏首字母，Nguyen 的文章在上述列表中排在第二位

顺序编码制、姓名编码制呈现单个文献的格式相同，区别在于：①参考文献列表的排序方式不同；②文章正文中的引文编号也随之变化。

你也许还希望为参考文献标注 PMID 号码。所有医学类的文章都有这个电子标签，支持在 PubMed 数据库中识别。操作方法是：将 PMID 写在文献条目的末尾，后面不加英文句号。

Christakos S. In search of regulatory circuits that control the biological activity of vitamin D. J Biol Chem. 2017 Oct 20;292(42):17559–17560. PMID: 29055009

美国国家医学图书馆创建的 PubMed 数据库涵盖了 4000 多份生物医学期刊，可以访问 http://www.ncbi.nlm.nih.gov/pubmed/ 进行查询。

10 AMA格式：生物医学、医学、护理学和牙医学

AMA 格式（表 10.1）主要用于生物医学、医学、护理学、牙科和生物学的一些相关领域，以牛津大学出版社 2007 年出版的 *AMA Manual of Style: A Guide for Authors and Editors*（第 10 版）为基础，线上版本 *AMA Manual of Style* 会在网站 www.amamanualofstyle.com 定期更新。❶

在文中，根据引文的出现顺序依次标注（1）、（2）、（3）……，这些序号同时指向文章末尾完整的参考文献列表。若文章或图书有多位合著者，最多允许列出六位。若超过六位，则只写出前三位作者，后面加上"et al."。无需将作者的名字拼写完整，用首字母代替即可，而且大写字母后面不需要英文句号和空格（例如：Lipson CH）。期刊的名称要用缩写，标准的缩写方式可查阅 PubMed 期刊数据库（http://www.ncbi.nlm.nih.gov/journals）。

表10.1 AMA格式

期刊论文	Neto AS, Schultz MJ. Optimizing the settings on the ventilator: high PEEP for all? *JAMA*. 2017;317(4):1413–1414. ▶ AMA 格式把参考文献的标题视为"句子"，"句首"的第一个字母大写，但是副标题的首字母无需大写 Drinka PJ, Krause PF, Nest LJ, Goodman BM. Determinants of vitamin D levels in nursing home residents. *J Am Med Dir Assoc*. 2007;8(2):76–79. ▶ 在 AMA 格式中，缩写的期刊名称中间不需要插入英文句号（上例 *Assoc* 后面的"."是用来分隔期刊名称和出版信息的） Ho EC, Parker JD, Austin PC, Tu JV, Wang X, Lee DS. Impact of nitrate use on survival in acute heart failure: a propensity-matched analysis. *Am Heart J*. 2016;5(2):e002531. https://doi.org/10.1161/JAHA.115.002531 Liu J, Xiong E, Zhu H, et al. Efficient induction of Ig gene hypermutation in ex-vivo-activated primary B cells. *J Immunol*. 2017;199(9)3023–3030. http://www.jimmunol.org/content/199/9/3023. Accessed October 5, 2017.

---

❶2020 年 AMA 格式发布了第 11 版。——译者注

	续表
期刊论文	▶ DOI 编码的应用日益普遍，事实上，现在很难找到不带有 DOI 的医学期刊了，因此非常推荐使用（记得补充 https://doi.org/）。如果标注的是普通网址，应同时写下访问日期；如果是 DOI，则不需要再列出网址或访问日期 ▶ 对于期刊论文的引用，AMA 格式建议从浏览器的地址栏复制网址，即使网址很长也没有关系，请不要引用主页地址 ▶ 在线期刊不一定有页码。如果可能，建议采用其他标志，比如文章编号或电子页号 ▶ 在作者的位置最多列出六位。若多于六位，就只写前三位，然后用 "et al."，下面的这篇论文有十五位共同作者，参考文献的写法是： Watson LM, Bamber E, Schnekenberg, et al. Dominant mutations in GRM1 cause spinocerebellar. *Am J Hum Genet*. 2017;101(4):638. https://doi.org/10.1016/j.ajhg.2017.09.006 ▶ 对于来自不同团队的作者（这在医学论文中很常见），用分号在中间隔开。你可以自行决定是否添加 "for" 或 "and" 等字样 Singh AK, Szczech L, Tang KL, et al; for CHOIR Investigators. Correction of anemia with epoetin alfa in chronic kidney disease. *N Engl J Med*. 2006;355(20):2085–2098. ▶ AMA 格式不需要为缩写添加句点，这条规则也适用于 "et al"（其中的 al 是缩写，大部分出版物要求写成 "et al."）
论文的摘要	Erkut S, Uckan S. Alveolar distraction osteogenesis and implant placement in a severely resorbed maxilla: a clinical report [abstract]. *J Prosthet Dent*. 2006; 95(5):340–343. Erkut S, Uckan S. Alveolar distraction osteogenesis and implant placement in a severely resorbed maxilla: a clinical report [abstract taken from *Dent Abstr*. 2007; 52(1):17–19]. *J Prosthet Dent*. 2006;95(5):340–343. ▶ 上例中的第一条是引用了文章自身的摘要，第二条则是发表在不同期刊上的同一份摘要
预印本或未发表的论文	Song Z. Using Medicare prices—toward equity and affordability in the ACA marketplace. [Epub ahead of print October 18, 2017]. *N Engl J Med*. https://doi.org/10.1056/NEJMp1710020 ▶ 若期刊正式出版，应为文献条目补充期数、页码信息 Flynn HW. Management options for vitreomacular traction: use an individualized approach. Paper presented at: Retina Subspecialty Day, Annual Meeting of the American Academy of Ophthalmology; October 14, 2016; Chicago, IL.
已发表的信件、评论或社论	Guazzi M, Reina G. Regarding article, Aspirin use and outcomes in a community-based cohort of 7352 patients discharged after first hospitalization for heart failure [letter]. *Circulation*. 2007;115(4):e54. https://doi.org/10.1161/CIRCULATIONAHA.106.646182. Baden LR, Rubin EJ, Morrissey S, Farrar JJ, Drazen JM. We can do better—improving outcomes in the midst of an emergency. [editorial]. *N Engl J Med*. 2017;377:1482–1484.
图书，单一作者	Basu S. *Modeling Public Health and Healthcare Systems*. New York, NY: Oxford University Press; 2017. ▶ 书名中的实词首字母大写。冠词首字母无需大写，除非它是标题的第一个词 Wiggins CE. *A Concise Guide to Orthopaedic and Musculoskeletal Impairment Ratings*. Philadelphia, PA: Lippincott Williams & Wilkins; 2007. Lizza JP. *Potentiality: Metaphysical and Bioethical Dimensions*. Baltimore, MD: Johns Hopkins University Press; 2014. ▶ 即使是知名城市，州的名称也要写出来

续表

图书，多位作者	Schlossberg DL, Samuel R. *Antibiotics Manual: A Guide to Commonly Used Antibiotics*. Hoboken, NJ: John Wiley & Sons; 2017.
图书，多个版本	Briggs GG, Freeman RK, Towers CV, Forinash AB. *Drugs in Pregnancy and Lactation: A Reference Guide to Fetal and Neonatal Risk*. 11th ed. Philadelphia, PA: Lippincott Williams & Wilkins; 2017. Mazze R, Strock ES, Simonson GD, Bergenstal RM. *Staged Diabetes Management*. 2nd rev ed. Hoboken, NJ: John Wiley & Sons; 2007. ▶ 版次写在书名和出版地之间。 ▶ 如果只是修订版（revised edition），没有指定版次，可在对应的位置写"Rev ed"（ed 后面加"."，用于隔开出版信息）
多卷本著作	Cone D, Brice JH, Delbridge TR, Myers B. *Emergency Medical Services: Clinical Practice and Systems Oversight*. 2 vols. Hoboken, NJ: John Wiley & Sons; 2015. ▶ 若只想引用第二卷： Cone D, Brice JH, Delbridge TR, Myers B. *Emergency Medical Services: Clinical Practice and Systems Oversight*. Vol 2. Hoboken, NJ: John Wiley & Sons; 2015. ▶ 请注意，"卷"（volume）的缩写后面没有句点。AMA 格式不需要为缩写添加句点 ▶ 如果各卷有独立于总标题的单独标题，你也可以根据某一卷书对应的作者（或编辑）、标题和出版日期来进行引用，然后标注总标题和所在卷号。整套书的编者可以不写 Chorghade MS, ed. *Drug Development*. Hoboken, NJ: John Wiley & Sons; 2007. *Drug Discovery and Development*; vol 2.
图书，多个版本，无作者	*Dorland's Illustrated Medical Dictionary*. 32nd ed. Philadelphia, PA: Saunders; 2011. *Nursing 2018 Drug Handbook*. 38th ed. Philadelphia, PA: Lippincott Williams & Wilkins; 2017.
主编图书	Hurlemann R, Grinevich V, eds. *Behavioral Pharmacology of Neuropeptides: Oxytocin*. New York, NY: Springer; 2018. Gravlee GP, Davis RF, Stammers AH, Ungerleider RM, eds. *Cardiopulmonary Bypass Principles and Practice*. 3rd ed. Philadelphia, PA: Lippincott Williams & Wilkins; 2007.
主编图书中的一章	Prinz J. Is the moral brain ever dispassionate? In: Decety J, Wheatley T. *The Moral Brain: A Multidisciplinary Perspective*. Cambridge, MA: MIT Press; 2015:51–68.
电子图书或在线图书	Dalal AK. *Cultural Psychology of Health in India: WellBeing, Medicine and Traditional Health Care*. New Delhi, India: Sage; 2016. http://web.a.ebscohost.com/ehost/ebookviewer/ebook/bmxlYmtfXzEyNTIzODZfX0FO0?sid=aad89e12-a6e0-494f-a52c-f7a4b4d7a72a@sessionmgr4006&vid=0&format=EB&rid=1. Accessed October 26, 2017. Yefenof E, ed. *Innate and Adaptive Immunity in the Tumor Microenvironment* [Kindle edition]. Vol 1. New York, NY: Springer; 2008.
政府文献	Global Task Force on Cholera Control. Ending cholera: a global roadmap to 2030. Geneva, Switzerland: World Health Organization; 2017. US Environmental Protection Agency, Office of Air and Radiation, Indoor Environments Division. *A Brief Guide to Mold, Moisture, and Your Home*. Washington (DC): Environmental Protection Agency; 2012. https://www.epa.gov/nscep. Modified September 2012. Accessed November 4, 2017. ▶ 如果是比较短的文章，就按照期刊的方式处理文件标题；如果是比较长的文件，就按照书籍的方式处理标题

续表

私人通信	▶ AMA 格式的参考文献列表不允许包含个人信件、邮件、短信、私人讨论、非正式谈话等私人通信。请在正文中合理引注
DVD 或线上视频	National Library of Medicine. *Changing the Face of Medicine* [DVD]. Washington, DC: Friends of the National Library of Medicine; 2004. Madara JL. Talk presented at: 2017 AMA annual meeting [video]. https://www.youtube.com/watch?v=d6-6UCiH728. Posted June 10, 2017. Accessed October 29, 2017.
数据库	RCSB Protein Data Bank. http://www.rcsb.org/pdb/home/home.do. Accessed May 5, 2017. National Institutes of Health. Office of Dietary Supplements. Dietary Supplements Subset Database. http://ods.od.nih.gov/Health_Information/IBIDS.aspx. Accessed June 30, 2016. National Center for Health Statistics. Tables of summary health statistics for the U.S. population. Atlanta, GA: Centers for Disease Control and Prevention; 2017. https://www.cdc.gov/nchs/nhis/shs/tables.htm. Updated January 24, 2017. Accessed October 26, 2017. *Vital Health Stat* 10 (243). DHHS Pub No 2010-1571. Hyattsville, MD: National Center for Health Statistics (US Dept of Health and Human Services); 2009.
网站、网页或博客	Zika cases in the United States: Cumulative Zika virus disease case counts in the United States, 2015–2017. Centers for Disease Control and Prevention website. https://www.cdc.gov/zika/reporting/case-counts.html. Updated October 26, 2017. Accessed November 22, 2017. Sun R, Karaca Z, Wong HS. Characteristics of homeless individuals using emergency department services in 2014. US Department of Health and Human Services website. https://www.hcup-us.ahrq.gov/reports/statbriefs/sb229-Homeless-ED-Visits-2014.jsp. Updated October 2017. Accessed November 3, 2017. Lamberts R. There will be patients like this. Musings of a Distractible Mind. http://more-distractible.org/musings/2017/10/3/there-will-be-patients-like-this. Published October 2, 2017. Accessed October 21, 2017.
播客	Whitaker R. Muscle groups in the thigh [podcast]. Instant Anatomy. October 13, 2017. https://www.instantanatomy.net/podcasts/IA009.mp3. Accessed December 1, 2017.

为了展示文内引注和参考文献的组合使用，我们以论文的开篇举例（表10.2）。

表10.2 AMA格式（参考文献排序举例）

论文开场白	This research deals with the ABC transporter family and builds on prior studies by Nguyen et al,[1] Sheps et al,[2] and Kerr.[3]
参考文献 （取决于在文中出现的次序）	1. Nguyen VNT, Moon S, Jung, K. Molecular biology: genome-wide expression analysis of rice ABC transporter family across spatio-temporal samples and in response to abiotic stresses. *J Plant Physiol*. 2014;171(14):1276–1288. 2. Sheps JA, Ralph S, Zhao Z, Baillie DL, Ling V. The ABC transporter gene family of Caenorhabditis elegans has implications for the evolutionary dynamics of multidrug resistance in eukaryotes. *Genome Biol*. 2004;5(3):R15. 3. Kerr ID. Sequence analysis of twin ATP binding cassette proteins involved in translational control, antibiotic resistance, and ribonuclease L inhibition. *Biochem Biophys Res Commun*. 2004;315(1):166–173. ▶ Nguyen 的文章最先被引用，因此排在首位。根据 AMA 格式的要求，"et al" 在句中也没有使用英文句号

最后一点说明：所有医学类的文章都有一种电子标签，即 PMID 号码。它不是文献的必需信息，但是列出 PMID 对读者和你自己回溯文献都有帮助。若在条目中包含 PMID 号码，则无需再填写 DOI 编码或网址。PMID 要写在文献的最后，然后加上英文句号。

Christakos S. In search of regulatory circuits that control the biological activity of vitamin D. *J Biol Chem*. 2017;292(42):17559–17560. PMID: 29055009.

文章的 PMID 号码可在 PubMed 数据库中识别，该数据库涵盖了几乎所有生物医学类的期刊。PubMed 数据库由美国国家医学图书馆创立，可以访问 http://www.ncbi.nlm.nih.gov/pubmed/ 查询详情。

11

ACS 格式：化学

美国化学学会（ACS）有自己的格式手册，具体可以参考 ACS 于 2006 年推出的 *The ACS Style Guide: Effective Communication of Scientific Information*（第 3 版），相同的内容也可在 ACS 的官网获取。总的来说，ACS 格式允许你采用下列体例：

- 文内引用可以仿照 APA 格式或 CSE 格式，写成作者-年份夹注。文末的参考文献列表（Reference list）按姓氏首字母排序。
- 若使用顺序编码制，参考文献列表依然位于文末，每个条目的编号取决于引文的出现顺序，文内引注有两种风格：
   - 上标数字，比如"[23]"。
   - 在圆括号内标记斜体数字，比如"(*23*)"。

每种体例都有大量的化学类期刊在使用。你的实验室、导师或目标期刊或许也有偏好的风格。一旦你选定了某一种，请在论文内部保持统一。

好消息是，无论采用哪种体例，需要的信息都是一样的。即使对于参考文献列表，每个条目的呈现方式也都一样，只是顺序可能不同而已。无论是数字编号还是字母顺序，都使用悬挂缩进，即条目的第一行维持正常长度，其余行全部缩进。

通过表 11.1，将看到：在 ACS 格式下引用图书和期刊论文有一些差异。例如，期刊的出版日期用加粗字体，而图书却不加粗。关于这些神秘的细节没有任何解释，我猜想也许是很久以前，化学家们被烟雾笼罩，然后莫名其妙就爱上了这样的规矩并沿用至今。

表 11.1　ACS 格式（化学）：参考文献和文内夹注

期刊论文	参考文献	Schoepff, L.; Kocher, L.; Durot S.; Heitz, V. Chemically Induced Breathing of Flexible Porphyrinic Covalent Cages. *J. Org. Chem.* **2017**, *82*, 5845–5851. ▶ 或 Schoepff, L.; Kocher, L.; Durot S.; Heitz, V. *J. Org. Chem.* **2017**, *82*, 5845–5851. Xing, Y.; Lin, H.; Wang, F.; Lu, P. An Efficient D-A Dyad for Solvent Polarity Sensor. *Sens. Actuators, B* **2006**, *114*, 28–31.

续表

期刊论文	参考文献	▶ 或 Xing, Y.; Lin, H.; Wang, F.; Lu, P. *Sens. Actuators, B* **2006**, *114*, 28–31. ▶ 作者姓名之间用分号隔开 ▶ 论文标题是可以省略的，但整个列表内部应保持一致。（此前，论文标题一直是省略的。但是 *ACS Style Guide* 现在认为文章的标题是必要的，因为它既能提示研究主题，还能帮助读者查找文献。）标题中的大小写按照书名的方式处理 ▶ 期刊名称用斜体，而且要缩写，缩写方式参照 *Chemical Abstracts Service Source Index*（CASSI）执行 ▶ 出版年份用加粗字体，卷号用斜体，页码用正常字体，无需添加 "pp" 的字样 Fadaei, E.; Martin-Arroyo, M.; Tafazzoli M.; Gonzalez-Rodriguez, D. Thermodynamic and Kinetic Stabilities of G-Quadruplexes in Apolar Solvents. *Org. Lett.* [Online] **2017**, *19*, 460–463. https://doi.org/10.1021/acs.orglett.6b03606 (accessed October 27, 2017). ▶ 与普通的网址相比，更推荐使用 DOI 编码（记得补充 https://doi.org/）
	文内夹注	(Schoepff et al., 2017) (Xing et al., 2006) (Fadaei, 2017)
化学文摘	参考文献	Taneda, A.; Shimizu, T.; Kawazoe, Y. *J. Phys.: Condens. Matter* **2001**, *13* (16), L305–312 (Eng.); *Chem. Abstr.* **2001**, *134*, 372018a. ▶ Taneda 等人的这篇文章已发表在期刊上，并被 *Chemical Abstracts* 引用。这条文献显示了文章全文和摘要的两个出处。摘要应排在第二位，用分号将其与文章分开 Taneda, A.; Shimizu, T.; Kawazoe, Y. *Chem. Abstr.* **2001**, *134*, 372018a. ▶ 上例呈现的是同一篇文章，但只提到了 *Chemical Abstracts* 这个来源。引用整篇论文会更好，然而前提是你必须确实读过完整的文章 *Chem. Abstr.* **2001**, *134*, 372018a. ▶ 依然是同一篇文章，此处只引用了 *Chemical Abstract* 的编号。"*134*, 372018a" 代表 CAS 收录号，134 是卷号，372018a 是该文摘在纸质版《化学文摘》中的摘要号。上述信息与 CAS 数据库内的信息对应，数据库已逐渐取代纸质版 ▶ 1967 年之前的 *Chemical Abstracts* 没有此类编号。引用时，要标注摘要在页面上的哪一栏。例如位于 1167 栏的摘要 f，可以写成 1167f（或 1167*f*） ▶ 最好还是像上一个版本那样，将 Taneda 等人的名字写出来
	文内夹注	(Taneda et al., 2001) (*Chem. Abstr.*, 2001)
图书，单一作者	参考文献	Bahadori, A. *Waste Management in the Chemical and Petroleum Industries*; John Wiley & Sons: Chichester, UK, 2014; pp 23–42. ▶ 书名不可省略，而且一定要用斜体 ▶ 出版商名称写在地区前面 ▶ 出版年份不用粗体字 ▶ 页码范围前面要标注 "pp" 字样。另一种方式是，如果引用了完整的一章，可以不写页码，而是写章节（如 Chapter 3）
	文内夹注	(Bahadori, 2014)

续表

图书，多位作者	参考文献	Theodore, L.; Dupont, R. R.; Ganesan, K. *Unit Operations in Environmental Engineering*; John Wiley & Sons: Chichester, UK., 2017. ▶ 如何应对作者太多的情况呢？*ACS Style Guide* 觉得可以都写出来，不过实际上一些化学期刊允许列出十位，然后加分号以及"et al."
	文内夹注	(Theodore et al., 2017) ▶ 此处最多允许列出两个名字。如果有三位或以上作者，要使用"et al."。假设上例共有两位作者，则夹注的形式为： (Theodore and Dupont, 2017)
在线图书	参考文献	Olovsson, I. *Wonders of Water: The Hydrogen Bond in Action* [Online]; World Scientific: Hackensack, NJ, 2017. http://www.worldscientific.com/worldscibooks/10.1142/10684#t=toc (accessed November 10, 2017).
	文内夹注	(Olovsson, 2017)
图书，多个版本	参考文献	Tro, N. J.; Vincent, J. J.; Livingston, E. J. *Laboratory Manual for Chemistry: A Molecular Approach*, 4th ed.; Pearson Education: Columbus, OH, 2017. ▶ 如果是修订版（revised edition），可以将版次"4th ed."替换为"Rev. ed."
	文内夹注	(Tro et al., 2017)
图书，多个版本，无作者	参考文献	*Reagent Chemicals: Specifications and Procedures*, 11th ed.; American Chemical Society: Washington, DC, 2016. *McGraw-Hill Encyclopedia of Science and Technology*, 11th ed.; McGraw-Hill: New York, 2012; 20 vols.
	文内夹注	(*Reagent Chemicals*, 2016). (*McGraw-Hill*, 2012) ▶ 如需具体引用某一卷： (*McGraw-Hill*, Vol. 6, 2012)
多卷本著作	参考文献	*The Encyclopedia of Mass Spectrometry*; Gross, M. L.; Caprioli, R., Eds.; Elsevier Science: Oxford, 2007; Vol. 6. *Hyphenated Methods*; Niessen, W., Ed. Vol. 8. In *The Encyclopedia of Mass Spectrometry*; Gross, M. L.; Caprioli, R., Eds.; Elsevier Science: Oxford, 2007.
	文内夹注	(Gross and Caprioli, Vol. 6, 2007) (Niessen, Vol. 8, 2007)
主编图书	参考文献	*Applications of Molecular Modeling to Challenges in Clean Energy*; Fitzgerald, G.; Govind, N., Eds.; American Chemical Society: Washington, DC, 2017. ▶ 或 Fitzgerald, G.; Govind, N., Eds.; *Applications of Molecular Modeling to Challenges in Clean Energy*; American Chemical Society: Washington, DC, 2017. ▶ 若只有编者，没有作者，则编者信息可灵活置于书名的前、后。如果参考文献列表需要按字母顺序排列，那么推荐将编者写在最前面，便于排序
	文内夹注	(Fitzgerald and Govind, 2017) ▶ 此处的编者信息仿照书籍作者的格式和要求处理
主编图书中的一章	参考文献	Kolosa, A.; Maciejowska, I. In *Chemistry Education and Sustainability in the Global Age*; Chiu, MH., et al., Eds.; Springer: New York, NY, 2013; pp 15–26. ▶ 或

续表

主编图书中的一章	参考文献	Kolosa, A.; Maciejowska, I. Polish Education Reform and Resulting Change in the Process of Chemical Education. In *Chemistry Education and Sustainability in the Global Age*; Chiu, MH., et al., Eds.; Springer: New York, NY, 2013; pp 15–26. ▶ 可以选择不写出章节的标题，但整个文献列表内部应保持一致。章节的标题与论文标题的格式相同
	文内夹注	(Kolosa and Maciejowska, 2013)
会议论文	参考文献	Fleming, L. Ecological Public Health, Harmful Algal Blooms and Climate Change. Presented at the 17th International Conference on Harmful Algae, Santa Catarina, Brazil, October 2016; Keynote speech.
	文内夹注	(Fleming, 2016)
工具书或百科全书	参考文献	*Ullman's Encyclopedia of Industrial Chemistry*, 7th ed. [Online]; John Wiley & Sons: Hoboken, NJ, 2014. https://doi.org/10.1002/14356007 (accessed September 16, 2017). Vaidya, R.; López, G.; Lopez, J. A. Nanotechnology (Molecular). *Van Nostrand's Encyclopedia of Chemistry* [Online]; John Wiley & Sons: Hoboken, NJ, 2005. http://www3.interscience.wiley.com/cgi-bin/mrwhome/110498369/HOME (accessed May 5, 2010).
	文内夹注	(*Ullman's Encyclopedia*, 2014)
政府文献	参考文献	US Consumer Product Safety Commission. *School Chemistry Laboratory Safety Guide* (October 2006); DHHS (NIOSH) Publication No. 2007-107; National Institute for Occupational Safety and Health: Cincinnati, OH, 2007. Energy Department. Meeting of the Biomass Research and Development Technical Advisory Committee. *Fed. Regist* [Online] **2017**, *82* (205), 49538. https://www.gpo.gov/fdsys/pkg/FR-2017-10-25/pdf/2017-23156.pdf (accessed October 2, 2017). National Emission Standards for Hazardous Air Pollutants for Source Categories from Oil and Natural Gas Production Facilities. *Fed. Regist.* **2007**, *72* (1), 26–43. ▶ 上例中的 *Federal Register* 被视为期刊
	文内夹注	(US Consumer Product Safety Commission, 2007) (Energy Department, 2017)
专利	参考文献	Adams, G. W.; Mullaney, J. S.; Oar, M. A. Hybrid Thermoplastic Gels and Their Methods of Making. US Patent 9,736,957, July 12, 2017. ▶ 专利名称允许省略
	文内夹注	(Adams et al., 2017)
视频	参考文献	Luceigh, B. A. *Chem TV: Organic Chemistry* 3.0 [CD-ROM]; Jones and Bartlett: Sudbury, MA, 2004. Davis, R. B. *Foundations of Organic Chemistry* [DVD]; The Great Courses: Chantilly, VA, 2014. Szydlo, A. *The Magic of Chemistry* [Online video]; The Royal Institution: London, UK, June 4, 2014. https://www.youtube.com/watch?v=0g8lANs6zpQ (accessed June 15, 2017).
	文内夹注	(Luceigh, 2004) (Davis, 2014) (Szydlo, 2014)

续表

网站、网页	参考文献	Biochemical Periodic Tables. http://eawag-bbd.ethz.ch/periodic/ (accessed August 10, 2017). ▶ 如果网页上能找到作者的名字，应置于条目的最前面： Oxtoby, J. Biochemical Periodic Tables. http://...
	文内夹注	(Biochemical Periodic Tables, 2017) ▶ 如果找不到网页文件的发布日期，可以在此处列上访问日期

12

物理学、天体物理学和天文学的引用格式

12.1 物理学常用的 AIP 格式

物理学的引用规范主要基于两本手册，一本是 1990 年美国物理联合会（AIP）推出的 *AIP Style Manual*（第 4 版），另一本是斯普林格出版社 2003 年出版的 *AIP Physics Desk Reference*（第 3 版）。大多数物理学期刊的文内引注和参考文献列表都遵循顺序编码制。条目在参考文献列表中的编号与它们在正文中出现的顺序一致。

文内引注的数字编号既可上标，也可使用方括号，即 "99" 或 "[99]"。*AIP Style Manual* 以及 AIP 的官方期刊（例如 *Chaos*、*Low Temperature Physics*）采用的是数字上标方式。而 AIP 旗下的 *AIP Physics Desk Reference* 则倾向于使用方括号，与之类似的还有 *Physical Review E* 等来自美国物理学会（American Physical Society）的期刊。方式没有对错之分，但是需要你在正文和参考文献中保持一致。（还有一些物理学期刊采用著者-出版年制，参考文献按字母顺序排列，悬挂缩进式排版。）

无论采用哪种体例，需要的信息都是一样的。尤其是参考文献（Reference list）中的论文和预印本（物理学领域常用的学术交流方式），每个条目的呈现方式几乎一样，也很简洁，应列出的信息包括：作者（与作品扉页上的写法一致，如 M. Shochet and S. Nagel）、期刊名称缩写、加粗的期刊卷号、文章的起始页码、圆括号标记年份（表 12.1）。

表12.1　AIP 格式（物理）：参考文献

期刊论文	[1] O. Budriga and V. Florescu, Euro. Phys. J. D **41**, 205 (2007). [2] G. Battimelli, Nucl. Phys. B **256**–**257**, 74 (2014). ▶ 省略文章的标题。期刊名称用缩写，无需斜体 ▶ 出版物的卷号（或期数）和系列号应使用加粗字体。假设要引用 *Physical Letters B* 第 466 期第 415 页的一篇文章，则写作：Phys. Lett. B **466**, 415 (1999).

期刊论文	[3] B. P. Williams and P. Lougovski, New J. Phys. **19**, 043003 (2017). <https://doi.org/10.1088/1367-2630/aa65de>. ▶ 上例中的期刊只在网上发行，在文章页码的位置填入的是文章编号（Williams 和 Lougovski 的论文编号是 043003） ▶ Williams 和 Lougovski 这篇论文的 DOI 编码是 10.1088/1367-2630/aa65de，建议也标注在条目中（前面补充 https://doi.org/，方便读者访问）。与普通的网址相比，DOI 编码更稳定、可靠
预印本	[1] B. Schuetrumpf, W. Nazarewicz, preprint, arXiv:1710.00579 [nucl-th] (2017). <https://arxiv.org/pdf/1710.00579.pdf>. [2] F. Dai, et al., preprint, arXiv:1710.00076 [astro-ph] (2017). <https://arxiv.org/pdf/1710.00076.pdf>. To be published in AJ. [3] A. J. M. Medved, preprint, arXiv:hep-th/0301010v2 (2003). <http://arxiv.org/PS_cache/hep-th/pdf/0301/0301010v2.pdf>. Published in High Energy Phys. **5**, 008 (2003). <https://doi.org/10.1088/1126-6708/2003/05/008>. ▶ 与普通的网址相比，更推荐使用 DOI 编码（记得补充 https://doi.org/） ▶ nucl-th 代表 nuclear theory，astro-ph 代表 astrophysics
图书，单一作者	[1] Alexander Piel, *Plasma Physics: An Introduction to Laboratory, Space, and Fusion Plasmas* (Springer, Berlin, 2017).
图书，多位作者	[1] Laurent Baulieu, John Iliopoulos, and Roland Seneor, *From Classical to Quantum Fields* (Oxford University Press, Oxford, 2017). ▶ *AIP Style Manual* 建议最多列出三位作者。如果超出三位，则只写第一位，后面加 "*et al.*"（注意用斜体），比如：Laurent Baulieu, *et al.*, *From Classical to Quantum …*
图书，多个版本	[1] J. J. Sakurai and Jim Napolitano, *Modern Quantum Mechanics*, 2nd ed. (Cambridge University Press, New York, 2017).
多卷本著作	[1] J.-P. Françoise, G. L. Naber, and T. S. Tsun, editors, *Encyclopedia of Mathematical Physics*, 5 vols. (Academic Press / Elsevier, San Diego, CA, 2006).
主编图书	[1] Yatendra S. Chaudhary, editor, *Solar Fuel Generation* (CRC Press / Taylor & Francis, Boca Raton, FL, 2017).
主编图书中的一章	[1] Heinz Georg Schuster, in *Collective Dynamics of Nonlinear and Disordered Systems*, edited by G. Radons, W. Just, and P. Häussler (Springer, Berlin, 2005).
数据库	[1] National Institutes of Standards and Technology, Physics Laboratory, Physical Reference Data. <http://physics.nist.gov/PhysRefData/>.

12.2 天体物理学和天文学的引用格式

天文学和天体物理学不使用 AIP 或物理学的引文体例，也可以说，上述学科不采用某种单一的引用格式。不过大部分主要期刊的风格是相似的：在文内，使用作者-年份夹注，然后有一个按字母顺序排列的参考文献列表。参考文献（Reference list）的格式也遵循一些最普遍的规则，即：

- 悬挂缩进式排版。
- 不含加粗或斜体字。
- 不写出作者完整的名字，而是用首字母代替。
- 共同作者之间用符号"&"连接。
- 将出版日期写在作者后面，二者之间无需逗号隔开。
- 省略文章标题。
- 列出书名以及出版商信息。
- 缩写期刊名称，经常只保留几个首字母。
- 只写出文章的起始页。
- 参考文献条目的最后不用英文句号。

这两个学科没有出版过官方的格式手册，引用格式比较多变。我根据主流期刊的处理方法做了整理，用比较规范的格式为你提供一些案例。表 12.2 主要参考了 *Astronomy and Astrophysics* 和 *Astrophysical Journal* 两本期刊的要求，为保持格式统一，某些地方进行了微调。

表12.2　天文学和天体物理学的参考文献

期刊论文	Saffe, C., et al. 2017, A&A, 604, L4 Koepferil, C. M., & Robitaille, T. P. 2017, 849, 4 Stroman, T., Pohl, M., & Niemiec, J. 2009, ApJ, 706, 38 Khomenko, E., & Collados, M. 2009, A&A 506, L5, https://doi.org/10.1051/0004-6361/200913030 ▶ 目前，所有的天文学、天体物理学和物理学文章都可通过标准的科学数据库在线获取。为条目添加 DOI 编码或其他检索信息能有效帮助读者查找对应的文献。与普通的网址相比，更推荐使用 DOI 编码（记得补充 https://doi.org/）
多篇文章，同一作者	Panaitescu, A. 2005a, MNRAS, 363, 1409 Panaitescu, A. 2005b, MNRAS, 362, 921 Pe'er, A., Long, K., & Piergiorgio, C. 2017, ApJ, 846, 54 Pe'er, A., & Ryde, F. 2017, IJMPD, 26, 1730018-296 ▶ 由于出现了两篇 Panaitescu 的论文，故分别记为 2005a 和 2005b ▶ 两篇 Pe'er 的论文有不同的合著者，因此没有写成 2017a、2017b
预印本	Kuruvilla, J., & Porciani, C. 2017, preprint, arXiv:1710.09379 [astro-ph]
图书，单一作者	Hoyle, F. 2015, Home Is Where the Wind Blows: Chapters from a Cosmologist's Life (Sausalito, CA: University Science Books) Harwit, M. 2006, Astrophysical Concepts (New York: Springer Science)
图书，多位作者	Mathur, R. B., Singh, B. P., Pande, S. 2017, Carbon Nanomaterials: Synthesis, Structure, Properties and Applications (Boca Raton, FL: CRC Press / Taylor & Francis)
主编图书中的一章	Thomas, J. 2012, in Solar and Astrophysical Magnetohydrodynamic Flows, ed. K. Tsinganos (Dordrecht, Netherlands: Kluwer Academic), 39
丛书中的一册	Branch, D., Wheeler, J.C. 2017, Supernova Explosions (Berlin: Springer), Astronomy and Astrophysics Library Series

续表

丛书中的一章	Daly, P. N. 2006, in Astronomical Data Analysis Software Systems XV, ed. C. Gabriel, C. Arviset, D. Ponz, & E. Solano (San Francisco: ASP), ASP Conf. Ser. 351, 4
未发表的论文或学位论文	Kilbourne, H, et al. 2017, US CLIVAR Workshop Report, Rep. 2017-3, https://doi.org/10.5065/D6KP80KR Fiore, F., Guainazzi, M., & Grandi, P. 1999, Cookbook for BeppoSAX NFI Spectral Analysis, available from https://heasarc.gsfc.nasa.gov/docs/sax/abc/saxabc/saxabc.html Lee, E. J. 2017, Ph.D. Diss, University of California, Berkeley
网站	SkyView, the Internet's Virtual Telescope <http://skyview.gsfc.nasa.gov/>

物理学领域的研究人员经常引用未发表的研究成果，例如即将刊发的会议论文或仍在进行中的研究。这些论文被称为预印本（preprints）或电子预印本（e-prints），它们聚焦的是最前沿领域，收录在电子文献库中。除了一些主要的研究机构，大量的预印本都储存在 arXiv.org 平台上（http://arxiv.org/），它在世界各地都有镜像站点，其中的论文都非常容易获取和下载。然而，除非你熟练掌握学科的前沿知识，否则很难读懂文章的内容。

arXiv 文献库中的预印本按照所属领域分类，如物理、天体物理、数学、定量生物学等，每个领域下还有一些重要的子领域。论文先投递至对应的子领域，然后根据收稿时间进行编号。预印本的引用方式与期刊论文类似，无需写出文章标题。我们来举例说明：

Biswarup, P. 2017, preprint, arXiv:1710.09634 [nucl-ex] <https://arxiv.org/abs/1710.09634>

Brax, P., Cespedes, S., Davis, A. 2017, preprint, arXiv:1710.09818 [astro-ph.CO]

或

Brax, P., Cespedes, S., Davis, A., 2017, preprint (arXiv:1710.09818 [astro-ph.CO])

与论文本身的高深内容相比，arXiv 的文献分类系统要简单得多。以 Brax 等人的文章为例，它位于天体物理学档案库（astro-ph），属于宇宙学和非银河系天体物理学的主题类别（CO），于 2017 年 10 月提交（1710），是当月同类别下的第 9818 篇文章，因此记为：1710.09818 [astro-ph.CO]。

对于 Biswarup 的文章，我在文献条目中列出了网址，这样读者就能下载文章的 PDF 版本了，不过网址是可以省略的。领域内的专业人士都知道在哪里可以找到 arXiv 的预印本，无论是在主站还是在镜像网站，所以只要列出 ID 就足够了。

即使尚未正式发表，预印本依然要像期刊论文一样列入参考文献。"未发表"并不意味着你引用时可以不标注。

13

数学、计算机科学和工程学的引用格式

13.1 数学和计算机科学

数学和计算机科学领域的论文有多种引用格式。数学论文中的参考文献通常是由 LaTeX 文档排版系统生成的，它可以产生许多不同的样式。作为论文的作者，你可能不需要手动创建参考文献，但仍有必要检查它们的准确性。

大多数情况下，文内引注都是由方括号里的数字给出的。数字依次编号，从[1]开始，完整的文献列表按相应的顺序排列，放在文末。另一种方式是根据文献作者姓氏的首字母排序组成列表，那么文中首次使用的文献可能对应[23]，因为作者是 Simpson；文章用到的最后一条参考文献可能标记为[2]，因为作者是 Biel。这些学科的文内引注极少标记页码，如果一定要写明，格式可以是：[23, p. 14]。

一种不太常见的格式是在方括号内标注作者姓氏以及出版日期的缩写，不写文献编号。比如 Hirano 和 Porter 于 2009 年发表在 *Econometrica* 第 77 卷的一篇文章应该标注为[HiPo09]、[HP09]或[HP]。具体方法由你决定，但在完整的文献条目中要把缩写也列进去，方便读者对照识别。

大多数的图书、论文都按所属子领域进行了分类，而且在数学评论 MR 数据库（Mathematical Reviews Database）中有唯一对应标识。所以无论你用哪种引注体系，都可以在条目的最后，即日期或页码之后，列出文献的 MR 编号。MR 数据库可通过美国数学学会（American Mathematical Society）的网站 http://www.ams.org/mr-database 检索。

如果你引用的文献可在网络获取，则建议在 MR 编号前面写出文献的 DOI 编码。若没有 MR 编号，文献条目就以 DOI 收尾。请记得将 DOI 编码写成网址的形式，例如一篇文章的 DOI 是 10.1515/jgth-2014-0041，改成网址需添加 https://doi.org/，即按照 https://doi.org/10.1515/jgth-2014-0041 的形式写入参考文献。

接下来，我依旧会用几张表格来归纳数学学科的引用格式，并且选用了一种比较规范、标准的体例。许多数学期刊的引用风格其实相当多变。有的在一篇文章中使用顺序编码制，到下一篇就改用首字母排序了。为了让乐趣加倍，同一条规则在不同的文章里还能发生变化，同样一篇文献的作者在论文 A 中写作 R. Zimmer，在论文 B 中记为 Zimmer, R.——只要我足够努力，也许能找到论文 C，在引用时把他写成了 Bob Zimmer。出版日期呢，有的论文将它们放入圆括号，有的则没有。一些文献列表（Reference list）将论文的标题写成斜体，期刊名是正常字体；翻到下一个文献列表又看到完全相反的操作。期刊的卷数也有加粗、不加粗两种情况……坦率地讲，我认为这些都无关大局，只要你做到前后一致，而且你的老师或合作的出版社也同意就没问题了。

对于各种变体和一些稀奇古怪的风格，我将不作考虑。表 13.1 参考主流的数学、计算机期刊的排版样式，采用了统一的规则。

表13.1 数学：带数字编号的参考文献（或按字母排序）

期刊论文	[1] A. Obus, *The local lifting problem for A4*, Algebra Number Theory **10** (2016), pp. 1683–1693. [2] D. W. Smith and R. G. Sanfelice, *A hybrid feedback control strategy for autonomous waypoint transitioning and loitering of unmanned aerial vehicles*, Nonlinear Anal. Hybrid Syst. **26** (2017), pp. 115–136. [3] N. P. Strickland, *Gross-Hopkins duality*, Topology **39** (2000), pp. 1021–1033. [4] ---, *Common subbundles and intersections of divisors*, Algebr. Geom. Topol. **2** (2002), pp. 1061–1118. ▶ 方括号和数字编号统一放在页面左侧。文献条目按使用的顺序编号排列或按作者姓氏首字母排序。若首字母相同，则按出版年份由远及近的顺序排列 ▶ 如果同一位作者重复出现（且无新的合著者），则可用三条长破折号替代相同的作者名，后面加逗号（英文长破折号的长度与字母 m 的宽度相近，因此得名 em dash。如果你不清楚如何键入长破折号，可以直接用三条连字符） ▶ 期刊名称用缩写形式，但不是斜体。标准的缩写名称可访问美国数学学会网站 http://www.ams.org/msnhtml/serials.pdf 进行查阅 [#] C. Banks, M. Elder, and G. A. Willis, *Simple groups of automorphisms of trees determined by their actions on finite subtrees*, J. Group Theory **18** (2015), 235–261. https://doi.org/10.1515/jgth-2014-0041. MR 3318536. [#] M. Leonelli, E. Riccomagno, and J. Q. Smith, *A symbolic algebra for the computation of expected utilities in multiplicative influence diagrams*, Ann. Math Artif. Intell. **81** (2017), 273–313. https://doi.org/10.1007/s10472-017-9553-y.
预印本	[#] J. Glänzel and R. Unger, *Clustering by optimal subsets to describe environment interdependences*, Technische Universität Chemnitz, Fakultät für Mathematik (Germany), preprint (2017-3). Available at https://www.tu-chemnitz.de/mathematik/preprint/2017/PREPRINT_03.pdf. [#] S. Dasgupta and J. Voight, *Sylvester's Problem and Mock Heegner Points*, preprint (2017), submitted for publication. Available at https://people.ucsc.edu/~sdasgup2/ylvester.pdf. [#] L. Chen, *Algorithms for deforming and contracting simply connected discrete closed manifolds (III)*, preprint (2017). Available at https://arxiv.org/pdf/1710.09819.pdf.

续表

预印本	[#] A. Hoque and K. Chakraborty, *Pell-type equations and class number of the maximal real subfield of a cyclotomic field*, preprint (2017), to appear in Ramanujan J. ArXiv:1710.09760 [math.NT]. [#] G. Lyubeznik, *On Switala's Matlis duality*, preprint (2017). arXiv: 1705.00021v1 [math.AC]. ▶ 与物理学类似，arXiv 预印本文献库中有大量的数学类论文，并且非常容易获取。数学学科的网址是 http://arxiv.org/archive/math，你可以引用预印本的网址（上例中 Chen 的文章），只标注文章的 arXiv ID 当然也是允许的（上例中 Hoque 和 Chakraborty 的文章） [#] X. Sun, *Singular structure of harmonic maps to trees*, preprint (2001), published as *Regularity of harmonic maps to trees*, Amer. J. Math. **125**(2003), pp. 737–771. MR1993740 (2004j:58014).
其他未发表的论文	[#] A. Iserles and S. P. Nørsett, *From high oscillation to rapid approximation II: Expansions in polyharmonic eigenfunctions*, DAMTP Tech. Rep. 2006/NA07. Department of Applied Mathematics and Theoretical Physics, University of Cambridge, Cambridge, UK, 2006. Available at http://www.damtp.cam.ac.uk/user/na/NA_papers/NA2006_07.pdf. [#] M. Fang, *Data enabled algorithms and analytics of material structural change informatics*, PhD diss, University of Memphis, 2017. [#] R. Viator, *Analysis of Maxwell's equations in passive layered media*, IMA postdoc seminar, Minneapolis, MN, 2017.
图书，单一作者	[#] D. Eisenbud, *The Geometry of Syzygies: A Second Course in Commutative Algebra and Algebraic Geometry*, e-book, Springer, New York, 2005. [#] D. Kondrashov, *Quantifying Life: A Symbiosis of Computation, Mathematics, and Biology*, e-book, University of Chicago Press, Chicago, 2016. [#] M. A. Parthasarathy, *Practical Software Estimation: Function Point Methods for Insourced and Outsourced Projects*, Addison-Wesley Professional, Upper Saddle River, NJ, 2007. ▶ 注意观察书名与论文、章节标题格式的不同之处，书名中实词的首字母都要大写（论文或章节标题的大写类似于写"句子"）
图书，多位作者	[#] A. M. Mathai and H. J. Haubold, *Linear Algebra: A Course for Physicists and Engineers*, Birkhäuser, Boston, 2017. ▶ 若作者人数过多，则只列出第一位，然后加"et al."字样，如：A. M. Mathai et al., *Linear Algebra ...*
图书，多个版本	[#] D. Hughes-Hallett et al., *Calculus: Single and Multivariable*, enhanced e-text, 7th ed., John Wiley & Sons, Hoboken, NJ, 2017. [#] B. Korte and J. Vygen, *Combinatorial Optimization: Theory and Algorithms*, 3rd ed., Springer-Verlag, Berlin, 2006. MR2171734 (2006d:90001).
多卷本著作	[#] F. Dillen and L. C. A. Verstraelen (eds.), *Handbook of Differential Geometry*, vol. 2, Elsevier, Amsterdam, 2006. [#] D. Knuth, *The Art of Computer Programming*, 3rd ed., vol. 4, fasc. 3: *Generating All Combinations and Partitions*, Addison-Wesley Professional, Upper Saddle River, NJ, 2005. MR2251472. [#] J. Humpherys, T. J. Jarvis, and E. J. Evans, *Foundations of Applied Mathematics*, vol. 1, SIAM, Philadelphia, PA, 2017.

续表

主编图书	[#] B. J. Copeland, C. J. Posy, and O. Shagrir (eds.), *Computability: Turning*, Gödel, Church, and Beyond, MIT Press, Cambridge, 2013.
译著	[#] P. G. Darvas, *Symmetry*, trans. by D. R. Evans, Springer, New York, 2007.
主编图书中的一章	[#] I. Boglaev, *Uniform convergent monotone iterates for nonlinear parabolic reaction-diffusion systems*, in Boundary and Interior Layers, Computational and Asymptotic Methods BAIL, Z. Huang, M. Stynes, Z. Zhang, eds., Springer, New York, 2017, pp. 35–48. ▶ 将章节的标题视为句子，所以只有首字母和专有名词大写；将书名视为标题，大写的部分更多
多卷本著作中的一章	[#] W. E. Hart, *A stationary point convergence theory of evolutionary algorithms*, in Foundations of Genetic Algorithms 4, R. K. Belew and M. D. Vose, eds., Morgan Kaufmann, San Francisco, 1997, pp. 127–134.
软件	[#] T. G. Kolda et al., *APPSPACK (Asynchronous Parallel Pattern Search Package)*; ver. 5.0.1, 2007. Available at http://software.sandia.gov/appspack/version5.0/pageDownloads.html. [#] *Windows 10 Pro*, Microsoft, Redmond, WA, 2015. Available at https://www.microsoft.com/en-us/store/d/windows-10-pro/df77x4d43rkt/48DN. ▶ 如果没有作者（上例中的 Microsoft 程序），按字母排序时以软件名称为依据

13.2 数学论文的文本样式

最后，所有的数学论文（无论其引用格式如何）都有专门的样式来规范定理、证明等标准术语的呈现方式，这些术语之后的文字内容也有特定的样式（表 13.2）。

表13.2 数学论文的文本样式

数学术语	术语的样式	术语后面文字的样式
定理（theorem）	THEOREM 或 **Theorem**	斜体
引理（lemma）	LEMMA 或 **Lemma**	斜体
推论（corollary）	COROLLARY 或 **Corollary**	斜体
证明（proof）	*Proof*	正常字体，无需斜体
定义（definition）	*Definition*	正常字体，无需斜体
注、注解（note）	*Note*	正常字体，无需斜体
评论（remark）	*Remark*	正常字体，无需斜体
观察（observation）	*Observation*	正常字体，无需斜体
例（example）	*Example*	正常字体，无需斜体

如需更多详细资料，请参考 Ellen Swanson 所著的 *Mathematics into Type*，美国数学学会于 1999 年推出了 Arlene O'Sean 和 Antoinette Schleyer 针对该书的升级版。

13.3 计算机科学：在编程中引用源代码

除了引述文章、引用文本，你在自己的程序中加入他人的代码和算法时，也应该进行引用。诚信原则仍然有效：如果引用了他人的成果，你必须公开承认，并告知读者成果的出处。建议你在注释中列出如下信息：被引代码段的作者、代码段的版本或日期、哪里可以获取，以及你采用这段代码的日期。请务必明确被引代码的起始和结束位置，如有改动，也应做出说明。

上述规则有一个例外。如果使用的是众所周知的算法，则无需专门标记引用。

13.4 工程学的 IEEE 引用格式

工程师们常用三种引用格式。第一种是 APA 格式（见第 8 章）。另外两种分别出自两个专业协会：电气电子工程师学会(IEEE，读作 I triple E) 和美国土木工程师协会（ASCE）。IEEE 格式受到大部分工程师的青睐，本节将简要介绍最新版的 *IEEE Editorial Style Manual*（线上版本可访问 http://www.ieee.org/获取），而 ASCE 格式将在下一节讲解。

IEEE 发行数以百计的期刊和会议论文集，在电气工程和电子领域之外，还涉及计算机科学、生物工程、土木工程、航空航天工程等诸多工程学科。如果你不确定自己的学科是否要遵守 IEEE 格式，请问问你的老师或助教。

大多数 IEEE 期刊在文内和参考文献（Reference list）中采用顺序编码制，与医学（第 10 章）和物理学（第 12 章）的引注体系类似。条目在参考文献列表中的编号与它们在正文中出现的顺序一致，无需考虑首字母。文内引注和文献列表中的数字编号写在方括号内，如[99]。参考文献列表为悬挂缩进排版，方便读者清晰查看数字编号。

正文中的数字编号也有多种应用场景，比如最常见的: according to Carmine [18] … 若需要插入多篇文献，可借助逗号或连字符: in the earlier studies [3], [22]–[24] … IEEE（表 13.3）有别于其他格式的方面在于它允许在方括号内补充页码、公式等信息，如[23, pp. 18–23]或[24, Fig. 13]。等式、公式的编号一般要加圆括号表示，因此在文内标记为[24, eq. (22)]。你比较熟悉页码（单数 p./复数 pp.）、等式（单数 eq./复数 eqs.）的缩写，其他缩写可以查阅词典。如果找不到缩写形式，可用完整形式。

文献列表中的作者名字写成首字母缩写。若有多位作者，可以都列出来。论文标题、报刊名称等的大小写格式采用"句子"的处理方式，即首字母和专有名词大写。书名、期刊名则视为标题，实词大写。期刊名要简写，官方手册 *IEEE Editorial Style Manual* 中列出了 IEEE 期刊的标准缩写方式，其他的期刊可到 PubMed 期刊数据库（http://www.ncbi.nlm.nih.gov/journals）中查询。

表13.3　IEEE 格式（工程学）：参考文献

期刊论文	[1] B. Komarath, J. Sarma, and S. Sawlani. "Pebbling meets coloring: Reversible pebble game on trees," *JCSS*, vol. 91, pp. 33–41, Feb. 2018, https://doi.org/10.1016/j.jcss. 2017. 07.009. [2] M. Inui and N. Umezu, "Quad pillars and delta pillars: Algorithms for converting dexel models to polyhedral models," *J. Comput. Inf. Sci. Eng.*, vol. 17, no. 3, 031001, Feb. 2017, https://doi.org/10.1115/1.4034737. ▶ number、volume、page 等词以及月份都要用缩写（May、June、July 三个月份应完整写出） ▶ IEEE 格式更倾向于列出刊发的月份，而非期数信息。对大部分理工科期刊来说，期刊页码是全年连续编号的，所以允许省略期数 ▶ 一些仅在网络发表的文章不会有页码。上例中的第二篇文章发表在 *Journal of Computing and Information Science in Engineering* 上，031001 是该期刊使用的文献标识符，可用来代替页码 ▶ IEEE 格式引用在线资源的原则与 APA 格式一致。即使引用的是印刷版文献，只要能找到 DOI 编码，都建议标注 DOI。若没有 DOI，则可引用文章所在的网页（建议从浏览器的地址栏复制网址）
未发表的文章或论文	[#] J. Yu, "Photonics-assisted millimeter-wave wireless communication," *IEEE J. Quantum Electron.*, to be published. https://doi.org/10.1109/JQE.2017.2765742. ▶ 若文章已被期刊接收，即将刊出，则标注"to be published"字样。若只是投稿，且尚未被接收，则应删去期刊名，并在文章标题之后标注"submitted for publication"字样 [#] M. Cole, "Advanced nanoengineering towards the electronics for tomorrow," presented at the 12th Int. Conf. on Surfaces, Coatings and Nanostructured Materials (NANOSMAT), Paris, France, Sept. 2017. ▶ conference 以及 national、international 等词要用缩写，不过事件或活动的名称应该写完整
已发表的会议论文	[#] L. Xie and Z. Mo, "Decision system of urban rail transit line conference construction sequence," in *Proc. 2nd Int. Conf. Transportation Engineering*, Chengdu, China, July 25–27, 2009, pp. 820–825. ▶ IEEE 手册要求注明页码 ▶ 如上文所述，IEEE 手册允许使用一些缩写词。上例中的会议论文集全称为 *Proceedings of the Second International Conference on Transportation Engineering*（请注意文献条目中省略了冠词、介词）
图书，单一作者	[#] S. W. Ellingson, *Radio Systems Engineering*. Cambridge, UK: Cambridge Univ. Press, 2016, p. 92. ▶ 若非引用全书，则应标明页码
图书，多位作者	[#] L. T. Biegler, I. E. Grossmann, and A. W. Westerberg, *Systematic Methods of Chemical Process Design*. Upper Saddle River, NJ, USA: Prentice Hall, 1997. ▶ IEEE 格式允许列出所有作者。但如果人数过多，也可以只写出第一位，然后加上"et al."（如 L. T. Biegler et al., *Systematic Methods* ...）
图书，多个版本	[#] G. Rizzoni and J. Kearns, *Principles and Applications of Electrical Engineering*, 2nd ed. Columbus, OH, USA: McGraw-Hill Education, 2016.

	续表
多卷本著作	[#] M. Muste, et al., Eds., *Experimental Hydraulics: Methods, Instrumentation, Data Processing and Management*, 2 vols. Boca Raton, FL, USA: CRC Press / Taylor & Francis, 2017. ▶ IEEE 格式中编者的首字母大写，即 Ed.（或复数 Eds.），足见该领域对大写字母 E 的偏爱
主编图书中的一章	[#] A. Wada, Y. Huang, and V. Bertero, "Innovative strategies in earthquake engineering," in *Earthquake Engineering: From Engineering Seismology to Performance- Based Engineering*, Y. Bozorgnia and V. Bertero, Eds. Boca Raton, FL, USA: CRC Press, 2004, pp. 637–675
电子图书或在线图书	[#] D. Haskell, A. Pillay, and C. Steinhorn, Eds., *Model Theory, Algebra, and Geometry* (MSRI Publications 39) [Online]. Available: http://www.msri.org/publications/books/Book39/contents.html. [#] D. Keith, *A Case for Climate Engineering*. Cambridge, MA, USA: MIT Press, 2013 [Kindle version]. ▶ 对于标注了网址的线上资源，可以省去"Online"字样。如果引用的是 Kindle 版或其他版本的电子书，可在方括号内注明（如[Kindle version]）
专利	[#] C. Entsfellner and H. Heuermann, "Vectorial network analyser," US Patent 9,801,082, October 24, 2017.

13.5　工程学的 ASCE 引用格式

与 IEEE 格式不同，美国土木工程师协会引注文献时偏爱使用著者-出版年制。整体风格与 APA 格式相似（见第 8 章），有一些细节上的差别。具体的说明可在 ASCE 官网获取。

著者-出版年制对应的是按首字母排序的参考文献，列表的每个条目都应在正文中提及，文内夹注和参考文献必须前后呼应。我们先看举例：

Brenner, B. (2009a). *Bridginess: More of the civil engineering life*, ASCE Press, Reston, Va.

Brenner, B. (2009b). "Infrastructure at the end." *Leadersh. Manage. Eng.* 9(4), 205–206.

Brenner, B. (2015). *Too much information: Living the civil engineering life*, ASCE Press, Reston, Va., 60–64.

请翻开第 8 章开头的例证（三篇 C. Lipson 的著作），对比分析。相同点是：作者的名字用首字母缩写，年份放入圆括号。相同作者的文献按时间顺序排列，相同作者、相同年份的著作按标题首字母排序，并在年份后面添加 a、b 等字母编号。

不同点是：ASCE 格式在文章、章节标题两端使用双引号。与大部分理工科期刊类似，期刊名要用缩写。你可以到原文页面上查找期刊的缩写形式，或访问 PubMed 期刊数据库（http://www.ncbi.nlm.nih.gov/journals）进行搜索。还要注意每个缩写词的末尾都用了句点。

文内夹注为作者-年份，二者之间不用逗号，即（Brenner 2009b）。如需注明页码，则可插入逗号，不同的条目之间用分号隔开，例如：（Brenner 2009a; Brenner 2015, p. 62）。以上就是 ASCE 格式的注意事项，而你的主要任务依然是确保风格一致。若已经选定了目标期刊，建议你参照其中最新的一些文章样本完成排版。如果从样本以及下面的表格中都没能找到你想引用的著作类型，可以浏览本书的第 8 章，对照仿写。

为便于对比，表 13.4 大多数案例与 IEEE 格式选取的例证相同。

表13.4　ASCE 格式（工程学）：参考文献

期刊论文	Komarath, B., Sarma, J., and Sawlani, S. (2018). "Pebbling meets coloring: Reversible pebble game on trees." *JCSS*. 91, 33–41, https://doi.org/10.1016/j.jcss.2017.07.009. Inui, M., and Umezu, N. (2017). "Quad pillars and delta pillars: Algorithms for converting dexel models to polyhedral models." *J. Comput. Inf. Sci. Eng.* 17(3), 031001, https://doi.org/10.1115/1.4034737. ▶ ASCE 格式需要你列出期刊的卷号、期数 ▶ 一些仅在网络发表的文章不会有页码。上例中 Inui 和 Umezu 的文章发表在 *Journal of Computing and Information Science in Engineering* 上，031001 是该期刊使用的文献标识符，可用来代替页码 ▶ 关于网络资源的引用，ASCE 格式没有过多描述。鉴于它与 APA 格式的相似性，可以按照 APA 的处理方式执行（建议征得老师的同意）。即使引用的是印刷版文献，只要能找到 DOI 编码，都建议标注 DOI。若没有 DOI，则可引用文章所在的网页（建议从浏览器的地址栏复制网址）
未发表的文章或论文	Yu, J. (2017). "Photonics-assisted millimeter-wave wireless communication." *IEEE J. Quantum Electron.*, in press. https://doi.org/10.1109/JQE.2017.2765742. ▶ 文献列表允许写出所有作者。对于文内夹注，如果作者人数超过三位，可以只写第一位，然后标注"et al."，即(Yu et al. 2017) ▶ Yu 的这篇文章尚未正式刊发，如果引用的是 early access 版本，建议如实说明 ▶ ASCE 建议不要将未发表的文章、论文、报告等内容写入参考文献列表，但你可以在正文中标注引用，例如：According to A. Stradivari (unpublished manuscript, January 2018) …
已发表的会议论文	Xie, L., and Mo, Z. (2009). "Decision system of urban rail transit line construction sequence." *Proc., 2nd Int. Conf. on Transportation Engineering*, Southwest Jiantong University, Chengdu, China, July 25–27, https://doi.org/10.1061/41039(345)136. ▶ 与 IEEE 格式类似，ASCE 要求缩写会议论文集的标题。不同之处在于，全称中的"of the"用逗号代替，介词 on 也没有删去 ▶ 上述作品已结集出版，可以视为一本书。如果是从书中引用同一篇文章，则可将会议地点改为出版商（ASCE, Reston, Va），并用页码代替 DOI 编码： Xie, L., and Mo, Z. (2009). "Decision system of urban rail transit line construction sequence." *Proc., 2nd Int. Conf. on Transportation Engineering*, ASCE, Reston, Va., 820–825.
图书，单一作者	Ellingson, S. W. (2016). *Radio systems engineering*, Cambridge Univ. Press, Cambridge, UK.
图书，多位作者	Biegler, L. T., Grossmann, I. E., and Westerberg, A. W. (1997). *Systematic methods of chemical process design*, Prentice Hall, Upper Saddle River, N.J. ▶ 文献列表允许写出所有作者。对于文内夹注，如果作者人数超过三位，可以只写第一位，然后标注"et al."，即(Biegler et al. 1997)

续表

图书，多个版本	Rizzoni, G., and Kearns, J. (2016). *Principles and applications of electrical engineering*, 2nd ed., McGraw-Hill Education, Columbus, Ohio. ▶ ASCE 建议使用"传统的"州名缩写而不是邮政服务缩写（比如宾夕法尼亚州要写 Pa., 不能写 PA）。俄亥俄州 Ohio 没有传统缩写。要找到正确的缩写，请查阅字典或 *The Chicago Manual of Style*
多卷本著作	Muste, M., Aberle, J., Admiraal, D., Ettema, R., Garcia, M. H., Lyn, D., Nikora, V., Rennie, C., eds. (2017). *Experimental hydraulics: Methods, instrumentation, data processing and management*, 2 vols., CRC Press, Boca Raton, Fla.
主编图书中的一章	Wada, A., Huang, Y., and Bertero, V. (2004). "Innovative strategies in earthquake engineering." *Earthquake engineering: From engineering seismology to performance-based engineering*, Y. Bozorgnia and V. Bertero, eds., CRC Press, Boca Raton, Fla., 637–675.
电子图书或在线图书	Haskell, D., Pillay, A., and Steinhorn, C., eds. (2000). *Model theory, algebra, and geometry* (Online), MSRI Publications 39, http://www.msri.org/publications/books/Book39/contents.html. Keith, D. (2013). *A Case for Climate Engineering* (Kindle version), MIT Press, Cambridge, Mass. ▶ 对于标注了网址的线上资源，可以省去"Online"字样。如果引用的是 Kindle 版或其他版本的电子书，可在圆括号内注明，即(Kindle version)
专利	Entsfellner, C. and Heuermann, H. (2017). "Vectorial network analyser," US Patent 9,801,082, October 24.

14

各类引用格式常见问题解答

14.1　哪些内容需要引用?

（1）论文中引用的所有内容都要标注出来吗?

基本如此。论文参考的数据、权威观点都应进行引注。直接引用和转述的部分也应标明出处。如果你引用了个人邮件、短信、访谈或与老师的对话等私人通信内容，请注明。若你的大段论述都围绕某篇特定文章展开，此时你既可以多次引注同一篇文献；也可以直接在正文里加入一段明确的陈述，说明该文献发挥的基础性作用（但仍需适量标注引用）。

只有一个例外——广为人知的事实不必注明出处。例如，*Declaration of Independence*（《独立宣言》）签署于 1776 年 7 月 4 日，为此引用任何权威资料都是多此一举。然而倘若你接下来要讨论美国大陆会议的政治问题，各路权威都可以到注释中施展拳脚。

（2）一篇论文究竟需要多少篇参考文献呢?

要看具体情况，而且没有确切的数字。不过，深入、透彻的研究论文可能每一页都会用到几篇参考文献，甚至更多也能接受。如果连续几页都没有引注，恐怕是出了问题。极有可能是你忘记标注参考文献了，建议你仔细复查。

（3）一定要包含许多不同出处的文献吗?

与你选取的主题的复杂程度有关，也取决于你的研究深度和论文的篇幅。一个复杂或饱受争议的主题需要多样化的资料来支撑论述，这样才能既有事实依据，又有不同角度。反之，如果只是一篇小论文，而且话题简单，或许参考少数几个资料来源就够了。如果不清楚，可以问一下老师的要求，顺便还能请教老师到哪儿获取高质量的文献。

在任何情况下，不要把较长、较复杂的论文建立在两三篇参考文献之上，文献的质量再高也不行。你的论文应该不仅仅是为了诠释他人作品（除非论文主题是对该学者作品的专门分析），你的论文应该是一个独立的原创作品。"原创"的起点是综合利用多种来源，并确保引用的文献涵盖了当前争议性话题的各方面意见。

你当然不需要照单全收、随声附和。但是对于较长的论文和争论激烈的议题，你至少要向读者表明你已经调研过不同的观点，对不同的思路进行了评估，并且有能力对关键争议点做出回应。

也就是说，你的评述可以是正面的，也可以是负面的。你可以公开反对某篇文献的观点，也可以提醒读者关注与之对立的视角（在论文中这样写即可：For an alternative view, see …）。

14.2 如何应对引用中的难题？

（1）可以在注释中展开分析与讨论吗？

除了理工类学科，大部分的引用格式都允许你这么做。某些简短的见解如果直接写在正文里，可能会分散读者的注意力，所以可以充分利用论文的脚注或尾注。尽管这些讨论性的内容不是主体，你仍然要把它们视为论文的一个部分，认真编写。但要注意不能喧宾夺主，毕竟正文才是重头戏。

如果文内引注用的是著者-出版年制，即（Tarcov, 2017），添加注解时需要额外再加一层标注，通常是添加上标数字编号，写成（Tarcov, 2017）[3] 的形式。数字 3 指向的是第三条解释性文字。

如果是科技论文，又恰好采用顺序编码制，解释性的文字就无法跟上标数字对应了，因为上标数字必须指向文末的参考文献。如果你一定要添加一两条注解，建议用星号或其他标记。

（2）以芝加哥格式和 MLA 格式为代表的一些体例，要求对标题进行缩写，应该遵循怎样的规则？

标准做法是去掉标题开头的冠词，删去副标题（冒号之后的部分）以及其他不必要的词。原则是保留标题中最能体现作品主题的关键词。关键词未必是标题开头的几个词。如果标题本身就很短，也可以不做缩写处理。请看举例：

完整标题	*Guardians of Language: The Grammarian and Society in Late Antiquity*
	How to Write a BA Thesis
	"Situating Kingship within an Embryonic Frame of Masculinity in Early India"
	"Autonomous Pigs"
缩写标题	*Guardians of Language*
	BA Thesis
	"Situating Kingship"（根据论文的写作目的，还可能缩写成）
	"Embryonic Frame of Masculinity"
	"Autonomous Pigs"

新版 MLA 格式的规定有所调整。目前的要求是：保留标题的第一个名词以及它前面的所有形容词。如果标题的开头不含名词短语，可以只保留标题的第一个词，只要能在参考文献（Works Cited）中对照识别即可。所以同样的四部作品用于 MLA 格式的文内夹注，应该写作：

缩写标题（MLA 格式）　　　*Guardians*
　　　　　　　　　　　　　　How
　　　　　　　　　　　　　　"Situating"
　　　　　　　　　　　　　　"Autonomous Pigs"

（3）一些科技论文要求缩写期刊名称，哪里可以找到标准的缩写形式？

最简单的方法是查看原版论文的首页，通常就能集齐期刊的缩写和引用当前文章所需的全部信息。图书馆内的参考馆员有时也会搭建各种网站，汇总不同领域期刊的缩写名称。比较有名的是 All That JAS: Journal Abbreviation Sources 平台（网址为 http://www.abbreviations.com/jas.asp），该平台的创立者 Gerry McKiernan 是一位目录学家，他就职于爱荷华州立大学图书馆，担任科技图书管理员。另一处可用资源是美国国家医学图书馆的公开数据库（网址为 http://www.ncbi.nlm.nih.gov/journals），其中有数万种期刊，而且不仅局限于医学领域。

（4）文献 A 的论述中提到了其他作者的著作 B，可以引用文献 A 作为著作 B 的来源吗？

这种情况不在少数，而且它是我们不断延伸阅读，了解他人思想的重要途径。比如你正在读 E. L. Jones 的一本书，其中有一段有趣的引文来自 Adam Smith（亚当·斯密），对你而言，Smith 的观点比 Jones 的论述更加贴合写作目的，因此你希望引用 Smith 的这段话。这很正常，只是该如何进行引注呢？

有不同的方案供你选择。一种是顺着 Jones 的注释或参考文献追溯 Adam Smith 的原著，阅读相关的部分，然后从资料的源头（即 Smith 的著作）引用。此时的引注无需提及 Jones 的作品。亚当·斯密在领域中享有盛誉，只要你开始搜集研究资料，几乎一定会读到他的著作，所以也就没必要特地言明 Jones 的向导作用。但出于诚信的考虑，你必须找出 Smith 的原著并阅读对应段落。

规则很简单：只能引用那些你确实参考过，并且是在正常研究过程中会遇到的文献，不要引用某人说过的含糊不清的话或者从别处听来的二手信息。就上面的案例来说，你无需把几百页的 Adam Smith 全都读完，但务必要浏览与引文相关的页面。学术诚信的首条基本原则：当你宣称自己完成了某项工作，你实际上确实做了。

如果没时间亲自查看 Smith 的原著（或原著是用你无法阅读的语言写成的），还有第二种方案。可以在论文中这样注释：Smith, *Wealth of Nations*, 123, as discussed in Jones, *European Miracle*. 通常情况下，不需要注明它出自 Jones 书中的哪一页，如果你想注明也是可以的。文内夹注的形式也略有调整，但预期目的是一样的，建议写成：(Smith 123, qtd. in Jones)。

第二种方案也是诚实的。引用了 Smith 的观点，同时表明该信息取自 Jones 的作品。学术诚信的另一条重要原则在此体现：如果参考了他人成果，需要如实引注。既然是将 Jones 的作品作为 Smith 观点的来源，那就如实写出来。

另一种可能是，Jones 的论述引导你发现了一部罕见或相对陌生的作品——如果没有 Jones 的深入调研、广泛讨论，你也许永远不会知道这部文献。刚刚的两种方案依然有效。例如，Jones 引用了 Paul Rycaut 写于 1668 年的著作 *The Present State of the Ottoman Empire*，我不熟悉奥斯曼帝国（the Ottoman Empire），自然很难通过自己检索找出这本书。坦白讲，如果 Jones 不说，我根本不会知道它的存在。没见过 Rycaut 的原文，现在却要引用，诚实的做法是：(Rycaut 54, cited in Jones)。然而，假设我的专业方向是奥斯曼帝国历史，根据 Jones 的启发，Rycaut 的专著刚好可以作为一手资料加以引用。为确保诚信，我一定要翻开 Rycaut 的原著，读完对应段落。

遗憾的是，一些学者无视规则，强行越过了转引这一步。即使他们从未听说过 Rycaut，甚至连一页 Smith 都没读过，也要闭着眼把原著写进参考文献。结果，Jones 在引用的过程中出现了偏差，而偷懒照抄的学者重复犯错——这在学术界真实存在！其根源在于：引用方式不当，并谎称引文出自 Smith。

真正理解诚信原则比背诵规则更重要：

- 只引用你在正常研究过程中发现并实际使用过的文献。
- 公开、诚实地标注引文及其来源。

坚守这些原则，诚实做学问。

14.3 参考文献

（1）论文一定要有 Bibliography 或 Reference list 吗？

一定要，除非你的论文采用了芝加哥格式下的完整注释。根据第 6 章的讲解，引文首次出现时的完整注释（而非简略注释）为读者提供了文献的标题、出版商等全部信息，因此对 Bibliography 不做强制要求。（当然，你的老师可能仍会要求你提供参考文献列表，建议事先询问。）

其他各种格式都需要参考文献列表的原因也很简单：文内的标注过于简洁，不足以充分描述资料的来源。

（2）为撰写论文所阅读的各种背景资料都要列入参考文献吗？

取决于你对该资料的依赖程度以及论文的引用格式。MLA 格式、APA 格式以及理工科论文的参考文献要求仅列出实际引用的条目。芝加哥格式相对灵活，允许你在注释中提到正文未引用的作品。

我的建议如下：如果某份材料对你的写作主题至关重要，请务必检查论文正文，确保已在某处对它进行引用；相应地，无论是哪种引用格式的参考文献，这份材料都应写入文献列表。如果某篇背景读物在你的研究中并不重要，就不要担心引注的问题。

（3）参考文献会引发人们对一篇论文的质疑吗？

会。读者会浏览参考文献列表，分析文献的类型和质量。以下五种问题需要留意：

- 陈旧、过时的资料。
- 列出的文献从整体上体现某种偏见。
- 遗漏当前主题中的关键著作。
- 资料来源的质量或可信度存疑。
- 过度依赖一至两处资料来源。

这些并不是列表本身的问题，问题还是出自论文的写作环节，然后通过参考文献显现出来。

年代久远的材料能很好地服务于特定的写作目的，但对其他情况则会显得不合时宜。在不少人眼中，Gibbon 的 *Decline and Fall of the Roman Empire*（《罗马帝国衰亡史》）是有史以来最伟大的历史著作，但今天不会再有人将它作为研究古罗马或拜占庭帝国的主要依据。自 Gibbon 之后的两个世纪以来，已经浮现出太多令人印象深刻的研究成果。如果此刻你需要综述拜占庭或古罗马的研究进展，*Decline and Fall* 这部书就有些过时了，把它作为写作的出发点也会危害文章的质量。不过，如果写作主题是伟大的历史作品、十八世纪视角或人们对待拜占庭帝国的观念演变，Gibbon 的作品就不只是切题了，它是不可或缺的参考文献。

"陈旧"的概念一定是相对的。对于历史、文学和某些数学的子领域而言，十年乃至十五年前发表的作品仍具有时效性，这些领域的知识变化速率大体如此。对于遗传学等高速发展的学科，也许不到一年时间，作品就过时了。假如一篇新鲜出炉的分子遗传学论文还在引用 2010 年的文献（或者再近一点，2015 年的文献），这篇论文背后的整个项目都会招来严重质疑。无论身处哪个研究领域，你都应该参考最好的作品，并确保它们尚未被更好、更新的突破所取代。

偏见、遗漏关键作品和过度依赖少数资料来源，预示着不同的写作问题[1]。"偏见"

---

[1] Ralph Berry, *The Research Project: How to Write It* (London: Routledge, 2000), 108–9.

代表你只看到多面性中的一面。针对一个富有争议的问题，如果你的文献呈现一边倒的态势，就可能意味着偏见。遗漏权威性的著作不仅会削弱你的文章质量，还会令读者怀疑你是否真的了解当前议题。

上述问题的补救措施都是一样的。特别是对长篇、复杂的论文来说，你的阅读范围须覆盖研究主题的各类主要著作，同时要借助参考文献把它们体现出来。

不管论文的长度如何，你都应保证资料来源的质量和可信度。老师和助教在这方面更有经验，他们接触过大量文献，能为你提供有价值的建议。

14.4 直接引用

（1）如果引文的原作者还引用了另一位作者，如何在论文里处理引注呢？

假设你在论文中写道：

As Michael Mandelbaum observes, "the English writer G. K. Chesterton once said America is 'a country with the soul of a church.'"[99]

看得出来，此句需要引用 Mandelbaum，那么问题是：是否还要将 Chesterton 这段文字的出处写出来呢？答案是不需要。然而在某些情况下，多提供一点有关第二层引文的信息，对读者是有益的。此类内容不妨写到脚注（或尾注）里：

[99] Michael Mandelbaum, *Mission Failure: America and the World*, 1993–2014 (Oxford: Oxford University Press, 2016), 9. The quote is a paraphrase from Chesterton's *What I Saw in America* (New York: Dodd, Mead, 1922).

（2）原著是用西班牙语、法语写成的，而引文是我自己翻译成英语的，如何做好引注？

在引文后面或参考文献中注明"my translation"字样即可，而且不必为每条引文重复标注。你只需在第一条引文后面告知读者：所有的引文都是自己的译文，同时列出原著的文献信息就可以了。

在一些论文里，你或许希望既引用原文，又附上译文，这也没问题。二者没有固定的顺序，先写出一个版本，然后将另一个放入方括号（或圆括号），比如：

In Madame Pompadour's famous phrase, "Après nous, le déluge" (After us, the flood). As it turned out, she was right.

14.5 网络资源

（1）在一个博客中读到了想要引用的信息，可是，这篇博文也不是资料的源头，相关内容还链接到了其他网站。应该引用哪个网址呢？

最好是引用信息的初始来源。在上例中，应该引用网站，而非博客。为确保学术诚信，你需要访问原始网站，实地确认。

不幸的是，未必总是能找到资料的源头。一个博客外链到另一个博客，另一个博客又链接给第三个博客，你很快就迷失在链接迷宫中，无法确定哪一处才是最初来源。那又该如何应对呢？其一，如果不能判断哪个网站是信息的源头，就引用看起来最好的那个，同时写上：信息源自他处（或，该资料未注明信息来源）。例如：Blog A, based on information from website B。其二，如果追溯不到信息的源头，应谨慎考虑到底要不要在文中引用——你往往很难判别这条信息究竟是事实还是杜撰。

参考纸质版的文献也是如此：你应该尽量引用信息的初始来源，而不是二手来源。请记住，你必须亲自检验每一条文献的出处，不得直接粘贴他人文献列表的对应内容。

（2）需要在文献条目中注明网络资源的访问日期吗？

不一定。有些引用格式对此有要求，但大部分格式没有强制规定。建议你记下访问日期，以防万一。比如说，如果网站上找不到官方的发布日期或最近的修改日期，那么访问日期就能派上用场了——毕竟，你要为读者建立一定的时间框架，证明参考资料在网站上的可用时间。请到前面的章节中查看更多例证。另外，尽管格式手册不做要求，一些导师会需要学生提供访问日期，建议提前询问。

（3）知道网址（URL）的概念，但什么是DOI？哪个要用来引注？有必要把它们都列上吗？

DOI 的全称是 Digital Object Identifier（数字对象标识符），是一种在网络上识别和交换知识产权的国际化体系。DOI 编码通常是一串数字，如 10.1086/681095，标记在资料的开头或顶部。期刊论文就属于这个体系。

引用 DOI 的正确方式是在编码前面添加"https://doi.org/"，因此上例应标注为 https://doi.org/10.1086/681095，这样 DOI 就能像网址一样直接访问了。这是一种相对新潮的做法，某些格式手册还未及时跟进，但我已在书中统一❶。请你试着把整串内容输入浏览器的地址栏，看一看能否直达文章页面。

即使改变了一份文件的网址，其 DOI 保持不变，所以 DOI 优于传统的 URL。只要标注了文献条目的 DOI 编码，任何人都能追溯到条目的来源。URL 则不然，你或许也意识

---

❶ 作者的预判没有错，最新的 APA 格式第 7 版手册就增加了此项规定。——译者注

到了网址可能会变化甚至消失，这就降低了一条文献的价值。只不过，大多数普通网站没有 DOI，像 Wikipedia、nytimes.com、YouTube 等平台，你只能引用网址。

若二者都可用，现阶段的大部分格式手册都建议用 DOI 代替 URL，这意味着你不必把它们都列上。详情可以查阅前面章节的相关说明。

（4）其他的在线标识符，如 PMID，如何使用？

PMID 号码多用于医学、生物学期刊，全称是 PubMed Identifier。PubMed 数据库涵盖了几乎所有生物医学类的期刊以及一些预印本，其网址是 http://www.ncbi.nlm.nih.gov/pubmed/，该网站还为用户提供了新手教程。PubMed 数据库由美国国家医学图书馆下属的国家生物技术信息中心开发建立，具有巨大的学术价值。

其他专业领域也有自己的电子标识符。例如，物理学的 arXiv，这是一个大型的预印本文献库，按子领域分类归档。

本节提到的在线标识符均不是文献条目的必需信息，但是写出来有助于自己和未来读者回溯文献。

引用网络资源的建议：做笔记时，应记下：

- 网站或网页的地址或 DOI 编码（尽量用复制、粘贴的方式，避免抄写错误）。
- 网站或网页的名称或描述。
- 访问日期。

上文已经讲过 URL 和 DOI（以及其他类型的在线标识符）的区别，也分析了记录访问日期的必要性（不论最终是否写入文献条目），此处不再赘述。

写下网站的名称或描述是因为，一旦网址失效（这种事的确发生过），你可以借助这些信息重新搜索。

建议把极具价值的网站加入书签。同时，为同一篇论文贡献资料的不同网站可以归入独立的收藏夹，集中管理，收藏夹以论文标题命名；避免将这些网站混入长长的收藏栏，为后期整理造成困难。

最后，如果预感到被引素材可能随时会变更甚至在网上消失，不妨将它打印出来存档。这适用于博客帖文、社交网站上的评论、维基百科中的文章或在线新闻报道……它们可能会被重置、删除、停用或改写，而且不会事先告知读者，建议你有所防备。不过，如果你引用的是学术论文的在线版本，也许就不用备份了。大多数期刊对论文的修改和再发布有相当严格的规定，它们也会妥善归档过往的期刊。

（5）不是有文献管理软件来帮我处理这一切吗？

不完全如此。确实有一些知名的软件能够从图书馆目录和其他在线数据库中抓取文献信息，然后将它们编排成各种引用格式。

这听起来不难，但此类程序有很大的局限性。用它们来应付书籍和近些年的期刊论文几乎不会出错，因为这些资料的来源信息比较标准，便于下载。当你想引用一份更古老或更不寻常的资料时，它们就无能为力了。无论是实物档案中的一封信、YouTube 上的一段视频，还是一条短信、美术馆中的一幅画，或是早已停刊的报纸中的一篇文章——根本没有哪个数据库能提供关于这些项目的完整信息。

因此，假如你的论文要大量引用专著和学术期刊，或者说你主要从事物理、生物科学研究或社会科学领域的定量研究，文献管理软件能发挥比较大的作用。假如各种被引材料的差异很大、来源多样，难以在标准数据库中进行编码，比如人文学科和诠释性社会科学的参考文献，文献管理软件就不是很实用。

具体来说，像 EndNote、RefWorks 和 Zotero 这样的程序允许自建文献库，并可将它们直接插入你的论文中，实现多种引用格式的引文编排以及样式切换。这些程序既可对接标准化的数据库，还支持手动录入文献信息。

好消息是，此类软件正在连接越来越多的数据库。大多数的图书馆目录、书目数据库允许将信息导出到软件中，而且有更多出版商的在线目录也加入了这个行列。但如果你从未接触过类似软件，要花费一些时间摸索每种功能的细节。

而人文和社科领域的研究者需要投入更多精力。多样化的被引材料意味着额外的工作量，由于无法一键下载文献信息，你经常要从零开始，新建条目，录入各项信息。数据录入有一套结构化的流程。你必须按部就班地操作，仔细输入信息要素，避免错别字，并且严格使用程序要求的标点符号。例如，我用过的一个软件要求多名作者必须用分号隔开——不是逗号，是分号。

接下来，为论文插入参考文献。这一步依然需要细心核查，确保准确，即使是从数据库直接下载的条目也不应错过。如果原始资料的编码不正确，到这一步会显露无遗。我就亲眼见证过。参考柏拉图的作品时，我引用了一个比较新的版本，软件生成的文献将这本书的编辑放在了合著者的位置。众所周知，柏拉图没有合著者。如果软件的结论成立，他恐怕要再等 2500 年才能与这位编辑相遇……后来，我回到软件，帮助柏拉图恢复了他独著者的身份。

在开始使用软件之前，有必要知道自己的论文将采用哪种引用格式。如此一来，你会清楚每种类型的著作应该在软件中输入什么、略过什么（每条文献都有无数的细节要素等你填充）。你甚至能看出软件遗漏的要素并提前加以修正。

最后一点提醒：如果你所在的学校有自己的文献管理软件，而你未来可能去往另一所大学深造，应该咨询一下当前软件能否将自己的文献库导出。这些费尽辛苦收集起来的数据，对未来的学习生涯也是一笔宝贵财富。

随着时间的推移，软件肯定会变得越来越好，数据库也会越来越全面，但就目前的程度来看，这类软件仍有可能为用户造成负担。

致 谢

  这本书力图呈现的是大学中各个学科的引用规范和学术诚信原则，需要倾听不同领域专家的声音，每一次的修订我都非常重视这一点。

  在"加入实验室"一节，Tom Christianson、Nancy Schwartz、Paul Streileman（生物科学）、Vera Dragisich（化学）、Stuart Gazes（物理）等专家分享了各自的视角，他们都曾担任芝加哥大学实验项目的负责人。Diane Hermann 为"小组作业"提出了许多宝贵意见，多年以来她一直担任数学系主任。类似的还有生物学家 Michael LaBarbera，他协助审读了"学术诚信"的相关章节。通过咨询芝加哥大学的 Vincent Bertolini、Helma Dik、Peter White，普林斯顿大学的 Robert Kaster,我进一步完善了"语言课"的部分。在"荣誉行为守则"方面，诺特丹大学的 Susan Pratt Rosato 和乔治敦大学的 Keir Lieber 与我分享了见解。芝加哥大学出版社的 William S.Strong 和 Perry Cartwright 是版权与合同问题专家，他们根据自己的经验对剽窃行为给出了忠告。森林湖学院的 James Marquardt 为"课堂参与"一节贡献了实用的行动方案。物理学家 Thomas Rosenbaum，曾任芝加哥大学教务长，现在是加州理工学院的校长，他也对许多章节提供了反馈意见。在 GeorgeGavrilis 的建议下，这一版增加了第 3 章"优秀笔记养成秘籍"的内容。

  上述话题还融入了许多在校大学生的真实想法。我尤其要感谢 Erik Cameron、John Schuessler、Jonathan Grossberg、Jennifer London 等同学，他们对自己的求学经历进行了充分的反思。

  为了深入探讨"学术诚信"，我请教了每天都要面对此类问题的导师、辅导员和系主任。大部分争议通常由学校的学生处协调解决，在那里我也有幸得到了两位资深工作人员的帮助：SusanArt，她曾多年担任芝加哥大学的教导主任，以及副主任 Jean Treese，她同时还是学校入学教育项目的负责人。她们二位深知学术诚信的重要性，也切实了解学生面临的困难，为我带来许多启发。

  至于引用规范的章节，我要特别感谢 Susannah Engstrom——她为这一版细致整理了各种引用格式的新要求，并搜集了新的著作类型和例证。还要感谢 Russell David Harper 的独到见解，他是官方《芝加哥手册》以及杜拉宾式引用体例的主要修订者。如果没有 Jenni Fry、Gerald Rosenberg、Janet Dodd、Karen J. Patrias、Peggy Robinson、Sharon Jennings、Peggy Perkins 等人的帮助，就不会有前面的两版图书，感谢他们对各类引用格式提供的具体建议。

  感谢芝加哥大学出版社，与我合作的每一位编辑都全力以赴、尽职尽责。Linda Halvorson 见证了本书初版的诞生，她全方位地打磨每一处细节，并时时给予鼓励。Paul Schellinger 以他一贯的专业精神促成了这本书的第二版。尤其还要感谢 Mary Laur，她为每一版作品都付出了努力，作为第三版图书的责任编辑，她全程监督，并为改进文本提供了宝贵建议。

  上面提到的所有人都就职于校园里的不同岗位，无论是学术出版、实验室，还是院系、学生处，一个相同的目标将大家聚在一起——每个人都希望看到真正的学习和越来越多诚实的学术成果，这才是大学教育的核心。正是因为他们的慷慨帮助，这本书才得以完成。

索 引

DVD 或线上视频（DVD or online video）
 AMA 格式   124
白皮书（white paper）
 APA 格式   103
百科全书（encyclopedia）
 ACS 格式   129
 APA 格式   103
 MLA 格式   84
 芝加哥格式   61
报纸文章（newspaper article）
 APA 格式   101
 MLA 格式   83
 芝加哥格式   58
毕业论文（thesis）
 APA 格式   101–102
 MLA 格式   83
 芝加哥格式   58–59
表格（table）
 APA 格式   105
 MLA 格式   87
 芝加哥格式   66
播客（podcast）
 AMA 格式   124
 APA 格式   109
 CSE 格式   117–118
 MLA 格式   90–91
 芝加哥格式   71
博客（blog）
 AMA 格式   124
 APA 格式   108
 CSE 格式   117–118
 MLA 格式   90
 芝加哥格式   70

词典（dictionary）
 APA 格式   103
 MLA 格式   84
 芝加哥格式   61
档案材料（archival material）
 MLA 格式   83–84
 芝加哥格式   59–60
地图（map）
 APA 格式   105
 MLA 格式   87
 芝加哥格式   66
电视节目（television program）
 APA 格式   104
 MLA 格式   86
 芝加哥格式   64–65
电影（film）
 APA 格式   104–105
 MLA 格式   86–87
 芝加哥格式   65
电子论坛、邮件列表（electronic forum or mailing list）
 APA 格式   110
 MLA 格式   91
 芝加哥格式   73
电子游戏（video game）
 MLA 格式   90
 芝加哥格式   71
短信（text message）
 APA 格式   110
 MLA 格式   91
 芝加哥格式   72
多媒体应用（multimedia app）
 APA 格式   109
 MLA 格式   90

芝加哥格式	71	天体物理学和天文学	134
多篇文章，同一作者（articles, several by same author）		芝加哥格式	58–59
		期刊论文（journal article）	
APA 格式	97	ACS 格式	126–127
天体物理学和天文学	133	ACSE 格式	142
访谈（interview）		AIP 格式	132
APA 格式	104	AMA 格式	121–122
MLA 格式	85	APA 格式	100–101
芝加哥格式	63	CSE 格式	115,118
工具书（reference work）		IEEE 格式	140
ACS 格式	129	MLA 格式	82–83
古典著作（classical work）		数学	136
APA 格式	104	天体物理学和天文学	133
芝加哥格式	62	芝加哥格式	56–57
古兰经（Koran）		墙报论文（poster session）	
MLA 格式	84	APA 格式	101–102
芝加哥格式	62	曲线图（graph）	
APA 格式	103	APA 格式	105
广告（advertisement）		MLA 格式	87
APA 格式	106	芝加哥格式	66
MLA 格式	89	软件（software）	
芝加哥格式	68–69	数学	138
灰色文献（gray literature）		APA 格式	109
APA 格式	103	社交媒体（social media）	
会议论文（conference paper）		APA 格式	110
ACS 格式	129	MLA 格式	91
ACSE 格式	142	芝加哥格式	72
IEEE 格式	140	圣经（Bible）	
讲座（lecture）		APA 格式	103
芝加哥格式	67	MLA 格式	84
脸书（Facebook）		芝加哥格式	62
APA 格式	110	诗歌（poem）	
MLA 格式	91	MLA 格式	85
芝加哥格式	71–72	芝加哥格式	63–64
论文，未发表（paper, unpublished）		视频（video/ video clip）	
ACSE 格式	142	ACS 格式	129
AMA 格式	122	CSE 格式	117–118
APA 格式	101–102	APA 格式	108–109
MLA 格式	83	MLA 格式	90
IEEE 格式	140	芝加哥格式	70–71
数学	137	手稿集（manuscript collection）	

MLA 格式	83–84	单一作者	
书评（review）		ACS 格式	127
APA 格式	101	ACSE 格式	142
MLA 格式	83	AIP 格式	132
芝加哥格式	58	AMA 格式	122
数据库（database）		APA 格式	97
AIP 格式	132	CSE 格式	115,118
AMA 格式	124	IEEE 格式	140
CSE 格式	117–118	MLA 格式	79
数据库或数据集（database or data set）		数学	137
APA 格式	107	天体物理学和天文学	133
私人通信（personal communication）		芝加哥格式	54
AMA 格式	124	电子图书	
APA 格式	104	ACSE 格式	143
MLA 格式	85	AMA 格式	123
芝加哥格式	63	APA 格式	99
私信（direct message）		CSE 格式	116,118
APA 格式	110	IEEE 格式	141
MLA 格式	91	MLA 格式	81
芝加哥格式	72	芝加哥格式	54
图（figure）		多本图书，同一作者	
APA 格式	105	APA 格式	97
MLA 格式	87	MLA 格式	80
芝加哥格式	66	芝加哥格式	55–56
图表（chart）		多个版本	
APA 格式	105	ACS 格式	128
MLA 格式	87	ACSE 格式	143
芝加哥格式	66	AIP 格式	132
图书（book）		AMA 格式	123
ACS 格式	127–129	APA 格式	98
AMA 格式	122–123	CSE 格式	116,118
AIP 格式	132	IEEE 格式	140
APA 格式	97–100	MLA 格式	80
ASCE	142–143	数学	137
CSE 格式	115–116	芝加哥格式	53
IEEE	140–141	多卷本中的单卷作品	
MLA 格式	79–82	APA 格式	99
丛书中的一册		MLA 格式	81
天体物理学和天文学	133	芝加哥格式	52
丛书中的一章		多卷本中的一章	
天体物理学和天文学	134	数学	138

多卷本著作		AMA 格式	123
ACS 格式	128	APA 格式	99
ACSE 格式	143	IEEE 格式	141
AIP 格式	132	CSE 格式	116,118
AMA 格式	123	MLA 格式	81
APA 格式	99	芝加哥格式	52
IEEE 格式	141	早期版本的重印版	
MLA 格式	81	APA 格式	99
数学	137	MLA 格式	82
芝加哥格式	53	芝加哥格式	54
多位作者		主编图书	
ACS 格式	128	ACS 格式	128
ACSE 格式	142	AIP 格式	132
AIP 格式	132	AMA 格式	123
AMA 格式	123	APA 格式	98
APA 格式	97–98	CSE 格式	116,118
CSE 格式	116,118	MLA 格式	80
IEEE 格式	140	数学	138
MLA 格式	80	芝加哥格式	54
数学	137	主编图书中的一章	
天体物理学和天文学	133	ACS 格式	128–129
芝加哥格式	53	ACSE 格式	143
非英文著作		AIP 格式	132
APA 格式	99	AMA 格式	123
机构作者		APA 格式	100
APA 格式	98	CSE 格式	116,118
匿名作者		IEEE 格式	141
MLA 格式	81	MLA 格式	82
芝加哥格式	54	数学	138
无作者		天体物理学和天文学	133
APA 格式	98	芝加哥格式	56
MLA 格式	81	芝加哥格式	52–56
芝加哥格式	55	推特（Twitter）	
译著		APA 格式	110
APA 格式	99	APA 格式	110
MLA 格式	82	MLA 格式	91
数学	138	芝加哥格式	71–72
芝加哥格式	56	网站、网页（website or page）	
在线图书		ACS 格式	130
ACS 格式	128	AMA 格式	124
ACSE 格式	143	APA 格式	108

CSE 格式	117–118	AMA 格式	122
MLA 格式	89–90	APA 格式	102
天体物理学和天文学	134	CSE 格式	116
芝加哥格式	69–70	数学	136–137
舞蹈演出（dance performance）		天体物理学和天文学	133
MLA 格式	86	芝加哥格式	59
芝加哥格式	64	杂志文章（magazine article）	
戏剧（play）		APA 格式	101
MLA 格式	86	MLA 格式	83
芝加哥格式	64	芝加哥格式	58
信件、评论或社论，已发表（published letter, comment, or editorial）		摘要（abstract）	
		ACS 格式	127
AMA 格式	122	AMA 格式	122
学位论文（dissertation）		APA 格式	100
APA 格式	101–102	MLA 格式	82
MLA 格式	83	芝加哥格式	57
天体物理学和天文学	134	照片（photograph）	
芝加哥格式	58–59	APA 格式	105
研究或技术报告（technical and research reports）		MLA 格式	87
		芝加哥格式	66
APA 格式	103	照片墙（Instagram）	
演讲、报告或讲座（speech, lecture, or talk）		APA 格式	110
		MLA 格式	91
APA 格式	104	芝加哥格式	71–72
MLA 格式	85	诊断手册（diagnostic manual）	
芝加哥格式	62–63,67	APA 格式	107–108
艺术作品（artwork）		诊断性测试（diagnostic test）	
MLA 格式	87	APA 格式	107
芝加哥格式	65–66	政策文件（policy paper）	
音乐（music）		APA 格式	103
APA 格式	106	政府文献（government document）	
MLA 格式	88	ACS 格式	129
芝加哥格式	66–68	AMA 格式	123
邮件（email）		APA 格式	106
APA 格式	110	CSE 格式	116–118
MLA 格式	91	MLA 格式	89
芝加哥格式	72	芝加哥格式	69
有声书（audiobook）		专利（patent）	
MLA 格式	88–89	ACS 格式	129
芝加哥格式	67	ACSE 格式	143
预印本（preprint）		IEEE 格式	141
AIP 格式	132		